Peter Bocker

ISDN
The Integrated Services Digital Network
Concept, Methods, Systems

In collaboration with
G. Arndt, V. Frantzen, O. Fundneider, L. Hagenhaus,
H. J. Rothamel, L. Schweizer

Second Edition with 108 Figures and 21 Tables

Springer-Verlag

Berlin Heidelberg NewYork
London Paris Tokyo
Hong Kong Barcelona Budapest

Dr. rer. nat. Peter Bocker
Gerhard Arndt
Viktor Frantzen
Oswald Fundneider
Lutz Hagenhaus
Dr.-Ing. Hans Jörg Rothamel
Lutz Schweizer

Siemens AG
Public Communication Networks Group,
Central Laboratories
Hofmannstrasse 51, W-8000 München 70,
Federal Republic of Germany

This is a completely revised and international edition on the basis of
ISDN – Das diensteintegrierende digitale Nachrichtennetz,
3rd edition, Springer-Verlag 1990

ISBN 3-540-54819-X 2nd ed. Springer-Verlag Berlin Heidelberg New York
ISBN 0-387-54819-X 2nd ed. Springer-Verlag New York Berlin Heidelberg

ISBN 3-540-17446-X 1st ed. Springer-Verlag Berlin Heidelberg New York
ISBN 0-387-17446-X 1st ed. Springer-Verlag New York Berlin Heidelberg

Library of Congress Cataloging-in-Publication Data
Bocker, P., 1933- [ISDN, das diensteintegrierende digitale Nachrichtennetz. English] ISDN, the integrated services digital network: concepts, methods, systems/Peter Bocker; in collaboration with G. Arndt . . . [et al.]. — 2nd ed. p. cm. Translation of: ISDN, das diensteintegrierende digitale Nachrichtennetz. 3rd ed. Includes bibliographical references and index.
ISBN 3-540-54819-X (Berlin: alk. paper). — ISBN 0-387-54819-X (New York: alk. paper)
1. Integrated services digital networks. I. Arndt. G. (Gerhard) II. Title.
TK5103, 7B6313 1992 621.382 — dc20 91-45902

© Springer-Verlag Berlin Heidelberg 1992
Printed in Germany

Typesetting: Macmillan Ltd., India; Printing: Mercedes-Druck, Berlin;
Binding: Lüderitz & Bauer GmbH, Berlin.
61/3020-5 4 3 2 1 0 – Printed on acid-free paper.

Preface for the Second Edition

The Integrated Services Digital Network (ISDN), still just a vision a decade ago, has now become reality: almost all PTTs and telecommunication carriers in the industrialized world have already introduced ISDN or are intending to do so in the near future. The reason for this is the significant benefits that this network offers users, network operators and manufacturers alike:

– Users obtain additional services with a variety of supplementary services providing them with better and simpler ways of meeting their communication needs – particularly for non-voice services. The ISDN subscriber access will also enable users to operate existing systems more economically than would be the case with various dedicated networks.
– The network operators benefit from the universal network concept which allows them to introduce new services and new supplementary services without large specific capital investment. Economic advantages will also flow from service integration, especially at subscriber access level, and from standardization of operation and maintenance for the range of services provided.
– Manufacturers welcome above all the standard network access, enabling one terminal to access several communication services, i.e. opening up a wider market.

The driving forces behind the ISDN are thus, firstly, cost-effective communication in various information types via the same subscriber access line and, secondly, sophisticated new services and supplementary services, especially in the increasingly important non-voice area.

This book, now in its second international edition, illustrates and describes the ISDN, its technology, its services and the principles on which ISDN terminals operate. The main focus of attention is those features of the ISDN which can be implemented using the cabling of the existing telephone network; this type of ISDN is generally referred to as narrowband ISDN or 64 kbit/s ISDN (although higher-capacity channels are also defined for this network). Communication between multimedia workstations (text, graphics, still and moving pictures, voice), fast data transmission (CAD/CAM), efficient interconnection of local area networks (LANs) and, in general, full-motion video communication in color, with a quality comparable to that of television pictures, will, however, only be possible when subscribers are connected via optical fibers to the broadband ISDN. Many details of the broadband ISDN are at present still under discussion. Consequently a comprehensive treatment of these issues was not yet aimed at in this book.

After analysing the role of communications, this book first discusses the services and supplementary services, which can be implemented in the ISDN. The structure of the network and the basic principles of network dimensioning are then illustrated. Much space is devoted to the subscriber access and associated user-network interfaces and signaling. The next section deals with the basic features of ISDN terminals. Switching in the ISDN is examined in detail, and the interworking of network nodes and user terminals is illustrated with the aid of examples, with particular reference to the special features of ISDN private branch exchanges. The transmission methods and transmission paths used in the ISDN are then described, and the last chapter of the book contains a critical evaluation of the ISDN for users in the office, at home, and en route. Finally, the current standards and recommendations relating to the ISDN are listed in an Annex.

The book is intended for engineers who need to be familiar with the requirements and features of the ISDN in order to design, construct or operate communication systems and it is also aimed at those engaged in teaching and research, as well as at a broader public with an interest in technical matters. It provides an overview of the background to this new development in communication and places its various features in context. In view of the international readership, the technical terms used in this book have been chosen so as to comply with international standards, in particular the International Electrotechnical Vocabulary IEV and relevant CCITT recommendations such as, G.701, I.112 and Q.9. When a term is used first, some national terms are given in brackets where appropriate.

In view of the rapid development of technology and progress in the area of standards, all sections of the book have been completely revised for its second international edition. Thus, the many new developments of recent years in the field of telecommunication services, switching and transmission have been taken into account. In particular the section entitled *Packet Switching Techniques in the ISDN* has been completely rewritten. This reflects the consolidated state of international discussion to date, as well as the significance of Asynchronous Transfer Mode (ATM) as base technology for broadband ISDN. The discussion of the hierarchy of digital transmission channels now also includes the Synchronous Digital Hierarchy (SDH). The numerous new or modified standards have been included in the Annex. Naturally all the literature references have also had to be updated.

As in the first edition, the general description of the ISDN concept and its implementation was prepared in collaboration with Dipl.-Ing. Gerhard Arndt, Dipl.-Ing. (FH) Lutz Hagenhaus, Dipl.-Ing. Oswald Fundneider, Dipl.-Ing. Viktor Frantzen, Dr.-Ing. Hans Jörg Rothamel and Dipl.-Ing. Lutz Schweizer.

Chapter 1 was written by P. Bocker, Chapter 2 by H.J. Rothamel; Chapter 3 by L. Hagenhaus; Sections 4.1 to 4.3 by O. Fundneider; Section 4.4 and Chapter 6 by V. Frantzen; Chapters 5 and 8 by G. Arndt; Chapter 7 and the Annex by

L. Schweizer. The book is also the product of much advice and information supplied by colleagues in the various development departments, both within and outside Siemens AG; I am particularly indebted to Dr.rer.nat. Rainer Händel and Dr.-Ing. Manfred Huber for their suggestions for updating Sections 4.1 to 4.3. and to L. Schweizer for his valuable help in reading the manuscript and suggesting amendments.

Thanks are due to my professional colleagues for all their support in producing this book. However, I would also again like to express my particular thanks to the above-named authors for their understanding and cooperation, which has enabled us once more to undertake the worthwhile task of presenting an international readership with this cohesive and updated text.

Munich, June 1991   P. Bocker

Contents

1 The Function of Telecommunications 1

1.1 Types of Communication 1
1.2 Information Types and Signals 1
1.3 Systems for Interactive Communication 3
1.4 Development of the Communication Networks. 3
    1.4.1 Existing Communication Networks. 3
    1.4.2 Integration of Services 4
    1.4.3 Flexibility and Network Intelligence 5
1.5 The ISDN Concept 6
1.6 The Integration of Services with Higher Bit Rates. 8
1.7 The Importance of Standardization. 9

2 Telecommunication Services 11

2.1 Definition of Services. 11
2.2 Definition of Service Attributes and Supplementary Services . . 17
2.3 Services in the ISDN with Bit Rates up to 64 kbit/s 19
    2.3.1 ISDN Services up to 64 kbit/s via B-Channels 19
        2.3.1.1 Conversational Bearer Services. 19
        2.3.1.2 Conversational Teleservices 19
        2.3.1.3 Messaging Services 22
        2.3.1.4 Retrieval Services (Teleservices). 23
        2.3.1.5 Distribution Services 23
    2.3.2 ISDN Services via the D-Channel 24
        2.3.2.1 Conversational Bearer Services with Low
               Throughput 24
        2.3.2.2 Conversational Teleservices 24
    2.3.3 Supplementary Services 24
    2.3.4 Existing Services from the Telephone Network
        in the ISDN 29
    2.3.5 Services in Conjunction with Terminals from Dedicated
        Text and Data Networks 30
2.4 Service and Network Interworking Facilities 31
    2.4.1 Intercommunication Within the ISDN 31

2.4.2 Intercommunication Using Network Interworking
      Facilities   32
2.5 Services with Higher Bit Rates   32

3 ISDN Structure   34

3.1 Network Organization   34
3.2 Initial Situation for the ISDN.   35
    3.2.1 Telephone Network   36
    3.2.2 Text and Data Networks   37
3.3 Functions of the ISDN Network Components   38
    3.3.1 Subscriber Lines in the ISDN.   38
    3.3.2 Local Exchanges in the ISDN   39
    3.3.3 Signaling Between ISDN Exchanges   39
    3.3.4 Operation and Maintenance in the ISDN   40
3.4 Network Dimensioning   40
    3.4.1 Basic Considerations   40
    3.4.2 Effects of Service Integration   42
          3.4.2.1 Holding Time, Traffic Intensity, Busy Hour Call
                  Attempts   42
          3.4.2.2 Shared Use of the Network Equipment. . . . . . · 44
    3.4.3 Traffic Routing   44
3.5 Interworking with Other Public Networks   46
    3.5.1 ISDN and the Analog Telephone Network   46
    3.5.2 ISDN and Public Data Networks   47
3.6 Interworking with Private Networks   48
3.7 Numbering   48
3.8 Implementation Strategies   49
    3.8.1 Overlay Network and Cell Approach   50
    3.8.2 Pragmatic Implementation Strategy   51
    3.8.3 Satellite Links in the ISDN   53
3.9 ISDN Implementation Schedules.   54

4 Subscriber Access   56

4.1 Configuration of the User Station   56
    4.1.1 Functional Groups of the User Station   56
    4.1.2 Network Termination NT1   58
    4.1.3 Network Termination NT2   59
    4.1.4 Terminal Adaptor TA   61
    4.1.5 Connecting Private Networks.   61
4.2 User-Network Interfaces   63
    4.2.1 Preliminary Remarks.   63

4.2.1.1 Channel Types   64
4.2.1.2 Access Types and Interface Structures   65
4.2.1.3 Operation of Traffic Channels   66
4.2.2 User-Network Interface for the Basic Access   67
4.2.2.1 Reference Configurations   67
4.2.2.2 Electrical Characteristics for Information
Transmission   69
4.2.2.3 D-Channel Access by Terminal Equipments . . .   70
4.2.2.4 Frame Structure   72
4.2.2.5 Activation and Deactivation.   74
4.2.2.6 Electrical Characteristics for Power Feeding . . .   75
4.2.3 User-Network Interface for the Primary Rate Access . .   76
4.3 User Signaling   78
4.3.1 Protocol Architecture.   79
4.3.2 Types of Connection   80
4.3.3 Special Features of ISDN Signaling   83
4.3.3.1 Functions of the Network Terminations NT1
and NT2   83
4.3.3.2 Call Establishment   83
4.3.3.3 Bus Configurations   86
4.3.3.4 Simultaneous Signaling Activities   87
4.3.4 Link Access Procedure on the D-Channel   88
4.3.4.1 Features of the Data Link Layer   88
4.3.4.2 Data Transfer Protocol   89
4.3.4.3 Assignment of Unique Terminal Endpoint
Identifiers   90
4.3.5 Signaling for Circuit-Switched Connections   91
4.3.5.1 Simple Call Establishment   91
4.3.5.2 Simple Call Clearing   93
4.3.5.3 Refined Call Establishment and Clearing   94
4.3.5.4 Control of Supplementary Services   95
4.3.5.5 User-User Signaling   95
4.3.5.6 Stimulus Protocol   95
4.4 Connection of Terminals with Conventional Interfaces
to the ISDN   99
4.4.1 ISDN Bearer Service and Public Data Network
Access Solutions   102
4.4.2 Connection of X.21 Terminal Equipment with Single-Step
Call Establishment   105
4.4.2.1 Mapping of the Call Establishment and
Clearing Procedures Between the X.21
and S Interfaces   105
4.4.2.2 Adaption Between the X.1 User Rates of X.21
Terminal Equipment and the ISDN
Information Transfer Rate of 64 kbit/s   105

4.4.2.3 Ready for Data Alignment in the 64 kbit/s
Channel Between the Terminal Adaptors
and Between the X.21 Terminal Equipments . . .   108
4.4.3 Connection of Data Terminal Equipment with V.-Series
Type Interfaces to the ISDN   108
4.4.3.1 Support of the Analog Tip/Ring Interface
in the ISDN   109
4.4.3.2 Support of V.-Series Interfaces in the ISDN . . .   109
4.4.4 Connection of Terminals with X.25 Interface to
the ISDN   110
4.4.4.1 Basic Characteristics   110
4.4.4.2 Point-to-Multipoint Signaling for Incoming
Virtual Calls   114
4.4.4.3 Access to Packet Switching via the B-Channel . .   115
4.4.4.4 Access to Packet Switching via the Signaling
Channel (D-Channel)   115
4.5 Packet Switching Techniques in ISDN   119
4.5.1 Frame Mode Bearer Services   122
4.5.2 Asynchronous Transfer Mode   123
4.5.3 Conclusion and Outlook   125

5 ISDN Terminals   127

5.1 Introductory Remarks   127
5.2 Basic Features of an ISDN Terminal   129
5.3 Single-Service Terminals Connected to the ISDN   130
5.3.1 ISDN Telephone   130
5.3.2 Terminals for Non-Voice Communication in the ISDN . .   132
5.4 Multi-Service Terminals   135
5.4.1 Multi-Service Terminals for Telephony and
Telematic Services   135
5.4.2 Multi-Service Terminals for Telephony and
Image Communication   137

6 Switching in the ISDN   138

6.1 Introduction   138
6.2 New Demands Placed on Switching Due to Service
Integration in the ISDN   146
6.2.1 Subscriber Access   146
6.2.2 Trunk Access   149
6.2.3 Switching Network   149
6.2.4 User-Network Signaling   153

6.2.5 Control 154
6.2.6 Interexchange Signaling 155
6.2.7 Operations, Administration and Maintenance 155
6.2.8 Timing and Network Synchronization 157
6.2.9 Interworking and Access to Special Equipment 158
6.3 Interexchange Signaling in the ISDN 158
6.3.1 Basic Characteristics of Interexchange Signaling
Using CCITT Signaling System No. 7 158
6.3.2 The Message Transfer Part (MTP) 161
6.3.3 Signaling Relations Between ISDN Exchanges 162
6.3.4 Protocol Architecture of ISDN Interexchange Signaling . 165
6.3.5 Implementation of ISDN Interexchange Signaling
in the Exchange 169
6.3.6 Intelligent Network 170
6.3.6.1 Flexible Service Control Architecture 170
6.3.6.2 Transaction Capabilities via CCITT Signaling
System No. 7 171
6.4 Corporate ISDN Networks 171
6.4.1 Fundamental Solutions 171
6.4.2 Structure and Features of an ISDN PABX 173
6.4.3 Tie Line Traffic Between ISDN PABXs 178

7 Transmission Methods in the ISDN 182

7.1 Introductory Remarks 182
7.2 The Hierarchy of Digital Transmission Channels 182
7.2.1 Basic Building Block: 64 kbit/s 182
7.2.2 Primary Multiplex Signals 184
7.2.3 Digital Multiplex Hierarchies 186
7.3 Transmission Media 187
7.3.1 Conductors in Cables 188
7.3.2 Radio Relay 189
7.4 Equipment for Transmitting Digital Signals on Cable
and Radio Links 190
7.4.1 General 190
7.4.2 Transmission on Cables in Trunk Circuits 192
7.4.3 Transmission on Subscriber Lines 193
7.4.4 Radio Relay Transmission 198
7.5 Multiplexed Signals and Multiplexing Equipment 198
7.5.1 Synchronous Multiplexed Signals 198
7.5.2 Digital Multiplexers 199
7.6 Network Synchronization 200
7.6.1 Necessity for Network Synchronization 200

7.6.2 Achieving Network Synchronization 201
7.6.3 Clock Supply Requirements 203
7.7 Disturbances and Transmission Performance 204
7.7.1 Effect of Bit Errors 204
7.7.2 Effect of Slips 206
7.7.3 Effect of Signal Delay 207
7.7.4 Effect of Jitter and Wander 208
7.7.5 Performance of 64 kbit/s Connection Types in
the ISDN 208

8 ISDN – The User's View 210

8.1 ISDN in the Office 210
8.1.1 Telephone Communication 210
8.1.2 Non-Voice Communication 211
8.1.2.1 Significance of Service Integration 211
8.1.2.2 Transmission Speed 213
8.1.2.3 Message Structure 214
8.2 ISDN in the Home 215
8.3 ISDN in Mobile Communication 218

Annex: CCITT Recommendations and Other Standards Relating
to the ISDN 220

References 237

Subject Index 251

1 The Function of Telecommunications

1.1 Types of Communication

Communication means the exchange of information. The function of telecommunications is to facilitate communication for people.

Communication is based on the exchange of information between two parties in an *interactive dialog*. Since the invention of telegraphy, electric current on wires and later high-frequency electromagnetic waves have been used to enable individuals to communicate over distance in a simple, convenient manner without physical transport of information media. People can thus send each other texts (telegraphy), talk to each other (telephony) and exchange drawings and illustrations (facsimile transmission, videotelephony) by electrical means. With the development of computers, the need also arose for information transfer between man and machine and from machine to machine (data transmission).

The advent of radio broadcasting also opened the way to *distributive communication*, i.e. communication between one transmitter and a number of receivers which can be reached without recourse to physical transport. Both freely propagating and guided electromagnetic waves are used for transmitting voice, sound and picture (sound broadcasting and television). Anyone who can be reached by these waves has immediate access to distributive systems of this kind.

Recently another type of communication has evolved, namely *retrieval communication*. In retrieval communication, which is being considered as a form of interactive communication (cf. Sect. 2.1), the user ascertains what is available from an information center and requests the selected information at a time chosen by him. He is not able to alter the contents of this information. Examples of this are the information services provided by videotex, broadcast teletext and full-channel broadcast videography [1.1]. The dialog is restricted to initial selection of the information to be retrieved.

1.2 Information Types and Signals

The information handled in communication systems is available in the form of the spoken or written word, as sound, picture or as information to be processed, i.e. "data". Equally varied are the electrical signals physically representing these information types, and their demands on the communication system.

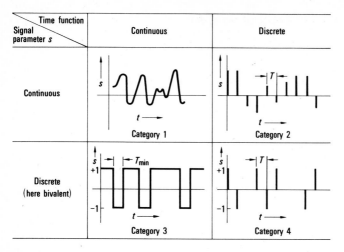

Fig. 1.1. Classification of Primary Signals. t Time

One of the characteristics of these signals is their *form of representation*. Their parameters may occur continuously or only in the form of discrete values, and they may have a relevant value at all times ("time-continuous") or only at certain discrete moments (Fig. 1.1). Technological progress has made it increasingly desirable for communication systems to handle signals that are discrete both in time and value; for instance, discretely timed signal sequences can be interleaved for transmission on a time-division multiplex basis (cf. Sect. 7.2). The primary signals mentioned are therefore converted more and more frequently into *binary, isochronous* signals.

Aside from the form of representation, the other important signal characteristic is the *quantity of information* to be transported in a unit of time. This determines either the upper limit frequency at which signal components must be transmitted, or – for binary isochronous signals – the required bit rate. The upper limit frequencies agreed by the CCITT[1] and CCIR[2], the international standardizing bodies of the telecommunications carriers, are 3400 Hz for telephony, 15 kHz for sound broadcasting and 5 MHz for television. The telephone channel thus defined corresponds to a bit rate of 64 kbit/s using pulse code modulation (cf. Sect. 7.2.1); moving images can also be transmitted via 64-kbit/s channels, albeit with reduced quality (see Sects. 2.3.1.2 and 5.4.2). TV-quality motion video requires bit rates of between 2 and 10 Mbit/s and above depending on the TV system and encoding method. The bit rates of text, data and still-image signals range widely from 50 bit/s through 64 kbit/s and upwards into the megabits-per-second region.

---

[1] Comité Consultatif International Télégraphique et Téléphonique
[2] Comité Consultatif International des Radiocommunications

Table 1.1. Systems for Interactive Communication (Examples)

Information type	Communication system
Speech	Telephony
Text	Telex
	Teletex
	Videotex (message service)
Data	Data transmission
Picture	Facsimile
	Telescript (telewriting)
	Videotex (information service)

1.3 Systems for Interactive Communication

A wide range of communication systems exists for the above mentioned information types: voice, text, data, image. Here we are primarily concerned with systems for interactive communication (including retrieval). Some examples are listed in Table 1.1. In each of these systems, communication can take place in different forms, e.g. between two or between several partners, without or with storage of information.

Each of the hitherto existing systems for interactive communication is designed to handle a single information type. This is in marked contrast to distributive communication, such as today's television service in which picture and sound can be transmitted together.

However, the ability to handle several information types in one system is also required for interactive communication: voice + moving picture, but also voice + fixed image (facsimile), voice + videotex, voice + data, text + data etc. The conventional communication networks, services and systems can only satisfy this requirement to a very limited extent.

1.4 Development of the Communication Networks

1.4.1 Existing Communication Networks

The switched networks currently available for interactive communication are primarily the following:

- The *telephone network*; this is the most widely available switched network for interactive communication; in addition to voice signals, it can also be used – with the aid of suitable additional equipment – to transmit still images (facsimile), data and telecontrol information as well as for data transfer between videotex computer centers [1.2].

- The *telex network*; this allows text to be transmitted between all the associated teleprinters in accordance with a standardized procedure [1.3].
- *Networks for text and data communication*; these employ various switching methods (circuit switching, packet switching; see Sect. 3.2.2) and have a range of user classes of service for which different data transmission rates as well as a variety of codes and speeds have been specified for the call setup and cleardown phases [1.4 to 1.6]. In Germany, the telex network has been expanded to form an independent integrated text and data network additionally incorporating a packet switching network [1.7, 1.8].

In addition to these switched networks, *leased lines* can also be provided for voice, text and data transmission.

1.4.2 Integration of Services

Examination of the existing state of telecommunications reveals that compatible communication networks and services on which the numerous different communication systems are operated are capable of providing communication between individual partners on a worldwide basis. However, interworking between the separate systems for voice, text, data and image communication is only possible in exceptional circumstances; both the terminals and the procedures for network access and communications handling are different and specifically geared to their particular task.

The further development of communication networks is therefore not merely a question of continually improving, extending and modernizing these systems, but also one of giving telecommunications a new direction; of moving away from the existing separate systems to a universal approach which will overcome the existing barriers between the different information types and technologies, the better to meet the primary function of telecommunications: to facilitate communication for people in all its aspects. In doing so, however, the possibility of interworking with existing networks must always be an attendant consideration.

Since for economic reasons telecommunications will be generally based on digital signals, a uniform digital communication network for all information types is the obvious solution.

Thus, this uniform network must therefore achieve two objectives. The first is the *integration of technologies*. On the one hand, the technologies used in the various services must converge so that e.g. voice and text signals can be transmitted by the same means on a subscriber line. On the other, the switching and transmission technologies within the network have to be integrated. For example, in a digitized network it is not necessary to fan out the time-division multiplexed signals before they reach the exchanges; these signals are fed directly to the switching equipment which then performs channel assignment in both time and space (see Sect. 6.2.3 and [1.9]).

The second objective for a uniform digital communication network is to introduce more advanced *new services* and to provide *service integration*. This means that

– terminal equipment for the different types of information can be connected to the network via uniform interfaces and procedures for call setup and cleardown,
– communication involving various types of information (voice + text, voice + image) can take place simultaneously or alternately via a single link.

1.4.3 Flexibility and Network Intelligence

In addition to network integration, a modern communication network must also offer every opportunity to the user of selecting the transmission rate and quality to suit his requirements. One step in this direction is being taken in the packet switching networks; in the future broadband networks, packet-oriented switching is being adopted as the general principle (Asynchronous Transfer Mode; see Sect. 4.5). In addition, a high degree of versatility is provided, enabling the user to communicate beyond the confines of the network, and allowing intercommunication between services (see Sect. 2.4); in some cases charges can be allocated to the called subscriber instead of the calling subscriber, as would otherwise be the case.

In addition to flexibility of this kind, modern communication networks should also provide facilities for message processing and store-and-forward; this is frequently termed "network intelligence" [1.10 to 1.12].

Network intelligence can provide *person-oriented* rather than traditional *station-oriented* communication. *Messaging services* with "mailbox functions" for voice, text, image or data can be established (see Sect. 2.3.1.3).

The user can also link up with *virtual private networks* within the public network, thereby obtaining for his business network the economic advantages, the possibilities of greater flexibility in tailoring services to individual requirements and the benefits of the greater availability of the public network to private facilities. Lastly, it should be possible for the user as well as for the operator not only to link up with the network compared to suit his requirements, i.e. changing services more intensively, but also to set up new services. For private service providers an "*open network access*" (Open Network Architecture ONA, Open Network Provision ONP [1.13, 1.14]) can be made available, via which they can offer their value-added services.

In order to ensure the necessary network and service quality, comprehensive operational control and supervision must be provided by the operator of the public network, but also by the operators of virtual private networks of this kind (Network Management [1.15]).

1.5 The ISDN Concept

The ISDN concept envisions service integration as well as the progressive convergence of technologies (see Sect. 1.4.2) [1.16, 1.17]. Moreover, it can be endowed – like the modern service-specific networks – with flexibility and network intelligence (see Sect. 1.4.3).

A concept as fundamental as this requires international consultations from an early stage (see Annex). It has been internationally agreed that the 64-kbit/s ISDN (Fig. 1.2) shall exhibit the following basic features:

- It is based on the digitized telephone network, in other words a network based on the digital 64 kbit/s telephone channel. The ISDN is therefore essentially a circuit switching network; however, packet-switched data traffic can also be handled in the ISDN (see Sects. 4.4.4 and 4.5).
- In the ISDN, connections from subscriber to subscriber are entirely digital.
- The basic access for a user (see Sect. 4.2.1.2) provides two 64 kbit/s basic channels (*B channels*) and one 16 kbit/s signaling channel (*D channel*) in each direction; the connections established over the two 64 kbit/s channels can be to different destinations. A primary rate access (see Sect. 4.2.3), mainly for connecting larger ISDN private branch exchanges, is also defined; this may comprise, depending on the multiplexing system used (see Sect. 7.2.3), up to 24 channels or up to thirty 64 kbit/s basic channels and one 64 kbit/s signaling channel. Basic and primary rate access can be provided on the copper wire pairs of existing subscriber lines (see Sect. 7.4.3). If optical fibers are used, it is also possible to specify ISDN access lines containing broadband channels, e.g. for video transmission.

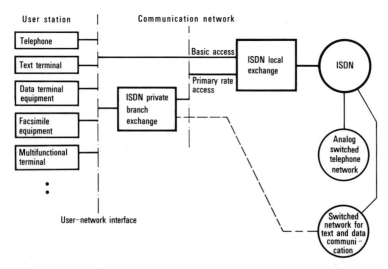

Fig. 1.2. The Integrated Services Digital Network ISDN

- To each subscriber a single directory number is allocated, irrespective of the number and type of voice, text, data and image communication services he requires.
- The universal user-network interface defined for the ISDN allows different terminals to be connected by a standard "communication socket" (Sect. 4.2). Accordingly, standard user procedures for call set-up and clear-down (Sect. 4.3) have been defined.
- The various terminals of a user station can be connected in a bus or star configuration (Sect. 4.1.3). The network not only establishes connections between the user stations (cf. Fig. 1.2) but additionally between those terminals within the user stations which are appropriate for the particular service required and are compatible (see Sect. 4.3.3).

Subscribers in existing networks, e.g. in the analog switched telephone network, are reached via the ISDN with the aid of interworking units (see Sect. 3.5). If no specific interworking unit is available, the subscriber terminal must be provided with a separate connection to the network in question (the dashed line in Fig. 1.2 shows how an ISDN private branch exchange is directly connected to the switched network for text and data communication).

The ISDN thus provides effective integration of technologies and of services for interactive communication.

As far as the *user* is concerned, the wide range of useful services and features available through the ISDN offers the following advantages:

- New and more versatile applications of communication due to the possibility of exchanging information of different types such as voice and text or text and data either simultaneously or alternately with one or more stations.
- Information exchange with the network through the high-capacity signaling channel even during an existing call without disturbing transmission of user information, e.g. for supplementary services such as *call waiting* or *advice of charge* (cf. Sect. 2.3.3).
- Improved accessibility of the user through the two basic channels and the facilities offered by the signaling channel and by change of service.
- Connection of terminals which can provide simple, uniform access not just to one but to many communication services supporting a variety of information types; even during an established connection, it is possible to effect a change of service including activation or deactivation of appropriate supplementary services (see Sect. 2.3.3).
- Disconnecting a terminal from a "communications socket" and reinserting it in another socket on the same user station without interrupting the existing connection.
- More widespread use of communication systems with high bit rates, mainly important for non-voice communication, e.g. facsimile (see Sect. 5.3) and data transmission.

The main advantages for the ISDN *carrier* are as follows:

- ISDN services and capabilities stimulate new applications and increased use of communication networks.
- A single, multi-purpose communication network with uniform technology for all services leads to uniform operation·and maintenance.
- The flexible digital principle on which the network is based enables new communication services to be introduced at comparatively low cost, even if only for trial purposes.

1.6 Integration of Services with Higher Bit Rates

In addition to voice, text, data and still-image communication via 64-kbit/s and higher-capacity channels (for approx. 2 Mbit/s), the ISDN concept also provides for communication services with higher bit rate requirements. For this purpose, the 64-kbit/s ISDN must evolve into a broadband ISDN, allowing a high degree of flexibility for fast data transmission, the creation of Metropolitan Area Networks (MANs), communication between multimedia workstations, video-conferencing, video telephony with TV quality, transmission and retrieval of video scenes. Aside from the individual communication services, broadband ISDN shall also provide distributive communication services, such as TV program distribution using existing TV standards as well as future high-definition television (HDTV). In addition, broadband ISDN must of course permit the connection of 64-kbit/s ISDN terminal equipment, as well as interoperation with the 64-kbit/s ISDN and other telephone and data networks [1.18].

The transmission method chosen for broadband ISDN is *asynchronous transfer mode* ATM (see Sect. 4.5), which allows the available bit rate to be adapted to suit arising requirements – a particularly important consideration in view of the wide variety of narrow and broadband applications and the absolute size of the bit rates. ATM allows links operating at any bit rate to be combined on a transmission section, thereby obviating the need to define high-speed channels at particular bit rates for particular services. As specified in CCITT Rec. 1.121 [1.19], ATM shall form the basis of the future broadband ISDN. As there is still a requirement to use existing equipment employing conventional synchronous circuit switching (synchronous transfer mode STM) and to provide cost-effective access to existing STM networks, STM will retain its importance as a transmission method alongside ATM, at least in the immediate future [1.20]. Traffic will then be gradually transferred from STM to ATM, until eventually the ATM network will become the universal network.

As discussions to finalize details concerning broadband interfaces and ATM parameters are still in progress, we shall not yet deal in depth with broadband ISDN in this book.

1.7 The Importance of Standardization

Developments in communications engineering are aimed at creating a future-proof, high-performance telecommunication infrastructure for the benefit of all, including the many small users with occasional, varied communication requirements, as well as the large professional users with their specialized needs. The basis of a telecommunications infrastructure of this kind are the public networks which provide the transport medium. These must be opened up to the user by means of services (see Sect. 2). As we take it for granted that we can nowadays make direct-dial telephone calls and send telexes or faxes worldwide, we can easily lose sight of the fundamental importance of the international standardization that has made such communication possible.

Standardization as it applies to ISDN is primarily concerned with international interfaces, i.e. with interexchange signaling and its special features in the ISDN (see Sect. 6.3), with the user-network interfaces (see Chapter 4) and with the ISDN services and supplementary service attributes (see Sect. 2.3). For these three areas, comprehensive and detailed standards have been agreed over recent years (see Annex), with increasing attention also being paid to the information processing aspects associated with the standardization of communication networks, systems and services.

Three standardizing organizations are basically involved worldwide (Fig. 1.3): the CCITT[1] which mainly deals with telecommunications, the ISO[2] and IEC[3] with the joint ISO/IEC technical committee JTC1[4] for the information technology area. In Europe, the corresponding organizations are ETSI[5], CEN[6] and CENELEC[7], and there are equivalent regional standardizing authorities in North America and Southeast Asia. Numerous activities must of

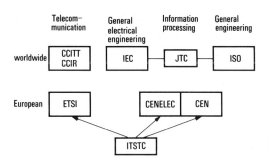

Fig. 1.3. Standardizing Organizations for the Communications Field

---

[1] Comité Consultatif International Télégraphique et Téléphonique
[2] International Organization for Standardization
[3] International Electrotechnical Commission
[4] ISO/IEC Joint Technical Committee 1 – Information Technology
[5] European Telecommunications Standards Institute
[6] European Committee for Standardization
[7] European Committee for Electrotechnical Standardization

course be coordinated, especially in the areas of overlap between the various bodies. For this purpose there are specific coordinating organizations, such as the ITSTC[1] in Europe, whose smooth functioning is acquiring increasing importance.

The standardization activities in the regional organizations, e.g. in the European Telecommunications Standards Institute ETSI, must be harmonized in content and timing with the corresponding activities worldwide. This is the only way to satisfy the requirements of users, operators and manufacturers alike for high-performance telecommunications on a global scale.

---

[1] Information Technology Steering Committee

To layer 7 are assigned the functions and protocols for controlling the applications (e.g. text transmission, text and data retrieval) and for editing and processing message contents for communication. The layer 7 functions also include the necessary evaluation of message-related information for controlling the communication process, such as the type of information, quality requirements, name or address of the communication partner or of the application process of a computer, authorizations and data protection (encryption).

Figure 2.2 shows by way of example the ISDN signaling protocols (D-channel) [2.2 to 2.7], and protocols of some teleservices for text and still image transmission [2.8 to 2.48] classified in accordance with the seven-layer structure.

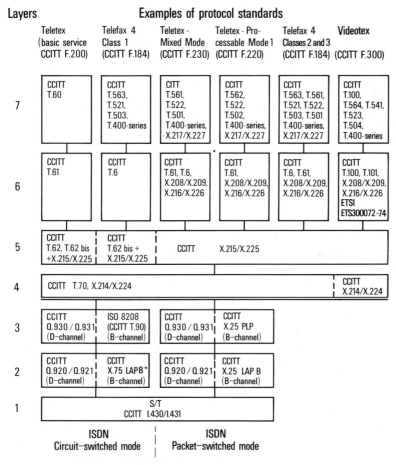

Fig. 2.2. CCITT, ISO and CEPT Recommendations for ISDN Signaling Protocols and Communication Protocols for Text and Still-Image Transmission in the ISDN.
LAP B Line Access Procedure B; * modified;
PLP Packet Layer Protocol

Depending on the extent of standardization of the communication functions and protocols, the services are subdivided into two groups by the CCITT: *bearer services* and *teleservices* [2.49 to 2.55].

Bearer services are used for e.g. unrestricted data and text transmission (clear channel) as implemented hitherto in circuit-switched and packet-switched data networks. The technical specifications of these services cover the transmission functions of OSI reference model layers 1 to 3 required for the transport of information. In circuit-switched bearer services, these are the functions assigned to layers 1 to 3 for signaling between user and network (ISDN: D-channel protocols) and the layer 1 functions (physical interface, transmission bit rate) for the transmission of user information (ISDN: layer 1 functions of the B-channel). In packet-switched bearer services the functions of layers 2 and 3 (transmission protocol, packet format) are also specified for transporting the user information. A bearer service only ensures the transport of information between the respective user-network interfaces. In other words the compatibility of the terminal equipment high layer protocols, in contrast to the teleservices, is the responsibility of the users of these terminals (Fig. 2.3a). The users may apply proprietary protocols or – in the future – they may use protocols standardized by organizations other than CCITT (in particular ISO).

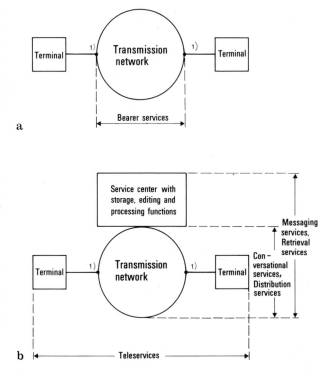

Fig. 2.3. a Sphere of Definition of the Bearer Services. **b** Spheres of Definition of the Teleservices.
[1]User-network interface

The *teleservices* are services for user-to-user and user-to-host communication, including specification of the communication functions of the terminals (Fig. 2.3b); these services include telephony, teletex, telefax and videotex. The communication functions comprise on the one hand all the transmission functions and communication protocols of layers 1 to 3, as mentioned for the bearer services. On the other hand, they include the functions and protocols for controlling the communication processes, if necessary for different information types (e.g. for transmission of alphanumeric characters or of picture elements (pixels) of a facsimile image), for communication oriented editing and processing of information at the transmit end, and for presenting the transmitted information on reproduction (output) at the receive end (layers 4 to 7). By virtue of their specifications, teleservices ensure compatibility between the terminals dedicated to the particular service, i.e. with respect to coding (character sets) and structuring (format) of the user information to be transmitted.

In addition to this OSI-oriented structuring a further classification of the services has been defined in CCITT covering ISDN services up to 64 kbit/s and the future broadband services. Depending on the different applications, communication forms and network resources required by the services, two main service categories have been identified: *Interactive services* and *distribution services* (Fig. 2.4). The interactive services are subdivided into three service classes, viz. the *conversational services*, the *messaging services* and the *retrieval services*. The distribution services contain services with and without user individual presentation control.

The interactive services and the distribution services may be standardized by the CCITT and offered by the public network carriers as teleservices and in some cases as bearer services.

Conversational services in general provide the means for bidirectional dialog communication with bidirectional, real-time (no store-and-forward) end-to-end

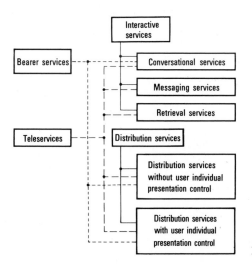

Fig. 2.4. Classification of Services

information transfer from user to user or between user and host (e.g. for data processing). The flow of user information may be bidirectional symmetric, bidirectional asymmetric and in some specific cases (e.g. such as surveillance) the flow of information may be undirectional. The information is generated by the sending user or users, and is dedicated to one or more individual communication partners at the receiving site. Examples of conversational services are telephony, audio conference, teletex, telefax, and data transmission. For those services, ISDN has to provide bidirectional connections with an unrestricted (transparent) bearer capability or a bearer capability dedicated to the type of information to be transported. Conference services require special conference facilities.

Messaging services offer user-to-user communication between individual users via storage units with store-and-forward, mailbox and/or message handling (e.g. information editing, processing and conversion) functions. Examples of messaging services are message handling services and mail services for audio information, text, data, graphics and high resolution images, such as voice mail, text mail, fax mail. These services require storage (messaging) units which may be provided centrally in the ISDN or in private networks.

The user of *retrieval services* is able to access information stored in information centers and generally provided for public use. This information will be sent to the user on demand only. The information can be retrieved on an individual basis, i.e. the time at which an information sequence is to start is under the control of the user. Examples are retrieval services for text, data, graphics, high resolution images and audio information, such as videotex. Retrieval services need connections with bearer capabilities dedicated to the types of information to be transported from the retrieval center to the user and to the user's request and selection of information in the backward direction. Retrieval centers may be installed in the ISDN (e.g. videotex center) or on the information provider's premises (e.g. as "external computer").

High layer functions of messaging and retrieval services, such as the input, storage and retrieval of information from centers, the mailbox functions and possible conversion functions for interface, protocol or bit rate adaptation, are also called "value added services" or "enhanced services" realized in special network nodes or external service centers as additions to the transport capabilities of a transmission network. These value added services are only communication functions and will be offered to the users as parts of telecommunication services, not as stand-alone services.

Distribution services with or without user-individual presentation control distribute a stream of information from a central source to an unlimited number of authorized receivers connected to the network. Distribution services *without* user-individual presentation control are the so-called broadcast services. They provide a continuous flow of information. The user can access this flow of information but cannot control the start and order of presentation of the broadcast information. As the user accesses the information at any point in time, it may not be presented from the beginning. Examples are broadcast services possibly for data and voice, and in the future broadband ISDN also for television and sound programs. To support those services, the network has to provide distribution nodes and/or unidirectional (broadcast) connections from the distribution node to the user, and, in most cases, connections from the user to the distribution node for program selection by the user.

Distribution services *with* user-individual presentation control provide an information stream structured as a sequence of information entities (e.g. frames) with cyclic repetition. The user can individually select the distributed information and can control its start and order of presentation. Due to the cyclic repetition, the information entity or entities selected by the user will always be presented from the beginning. One example of such services is broadcast videography or teletext. In general, the broadcast information is selected at the user's terminal, i.e. the selection is not the task of the network.

Layer 4 to 7 functions of the retrieval and messaging services, e.g. for information input, storage in and retrieval of information from text/data banks of retrieval service information centers and the mailbox functions of the messaging services, as well as the conversion functions for interface, protocol and bit rate adaption, are often referred to as "value-added services". However, these designations are misleading, as the so-called added-value services are only offered to users as communication functions within particular teleservices and bearer services and not as services in their own right. Functions of this kind can be implemented within a network in network notes equipped with special service modules or outside the network in external service centers.

The majority of standardized services used today are *single-medium* services, i.e. they are used for transmitting one type of information, such as voice, text or still image (graphics), etc. In future, however, more multimedia services are to be anticipated. Multimedia services offer the transmission of several types of information. For example, the audiovisual services "Videophone" and "Video-conferencing" mainly carry moving pictures and voice; multimedia document transfer and retrieval services (e.g. Videotex, Teletex mixed-mode) offer text, data, graphics and still images and, in the future, "mixed" documents containing also voice and moving picture information.

2.2 Definition of Service Attributes and Supplementary Services

As indicated in Fig. 2.5 and [2.56], the service attributes describing a communication service can be divided into:

● user related, technical service attributes
● carrier related, operational and commercial service attributes.

The user related service attributes can be further subdivided into *general access attributes*, *basic service attributes* and *supplementary services*. The general access attributes describe the ISDN features relating to the subscriber line circuit and

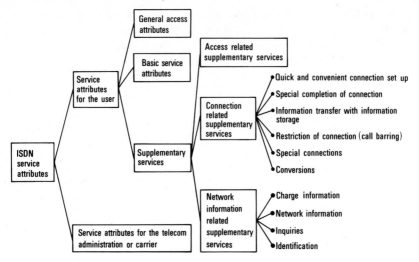

Fig. 2.5. Service Attributes and Supplementary Services in the ISDN

usable for all services, such as the characteristics of the user connection device, the type of connection (*switched or permanent*) and operating modes of the ISDN access (*multiservice operation, change of service*).

The basic service attributes describe a service in its basic configuration (minimum configuration [2.8–2.12], service attributes [2.49–2.55]). These are the information transport attributes – type of switching, bit rate, symmetry of information flow, communication establishment and configuration –, the network access features – transmission channels and protocols used (in layers 1 to 3) –, the functions and protocols of layers 4-7 – types of information, character sets, graphics representation, image and voice supported by the service, etc. – as well as service quality and intercommunication features.

The *supplementary services* modify or augment the basic attributes and characteristics of a service, e.g. in terms of higher communication quality and enhanced communication convenience. They are not offered to a user separately as stand-alone services but only in conjunction with the basic service attributes of a communication service. The same supplementary service can be used with different services. Examples of this are *abbreviated dialing, call waiting, closed user group*. The main supplementary services in the ISDN are described more fully in Sect. 2.3.3.

Corresponding to their assignment to the lower layers 1 to 3 or the higher layers 4 to 7 of the OSI reference model, the service attributes are designated low layer attributes describing access to the network, or high layer attributes concerned with communication between terminals (end-to-end) and oriented towards applications [2.49]. The basic service attributes and supplementary services are implemented in the communication networks and/or terminals with the aid of appropriate functional elements. Here, too, the same functional element, e.g. *transmission of the subscriber line identification, terminal compatibility check* or *access authorization check*, can be used for several basic service attributes and supplementary services.

2.3 Services in the ISDN with Bit Rates up to 64 kbit/s

Table 2.1 gives a list of ISDN services. In the table, each group of services, namely the services via B-channels or via the D-channel, the existing services from the telephone network and some data services of digital data networks as well as the services for connecting existing terminal equipment, are arranged according to the classification given in Sect. 2.1. The services are described in detail below. The status of implementation of ISDN services and plans for future introduction of services in the ISDN in several countries are given in [2.56 to 2.62].

2.3.1 ISDN Services up to 64 kbit/s via B-Channels

The ISDN 64 kbit/s channel connections between any users allow ISDN services to be introduced which are optimized for this transmission rate [2.52, 2.53, 2.55, 2.60, 2.63]. (For an interim period some ISDNs or parts thereof may only be capable of supporting 56 kbit/s unrestricted bearer services, cf. Sect. 7.2.3). The standardized signaling (D-channel) protocol and the uniform transmission bit rate on the traffic channel (B-channel) will also facilitate rapid and cost-effective implementation of new services and terminals.

2.3.1.1 Conversational Bearer Services

The powerful 64 kbit/s ISDN transmission channels provide sufficient transmission capacity for most data transmission applications in the office and the home. Data transmission services established on these communication paths using both switched and permanent connections open up new applications in the business and private sectors, far exceeding the capabilities of bearer services with different transmission speeds in existing dedicated circuit and packet switched data networks. Example of this range of services are the 64 kbit/s unrestricted bearer services standardized by CCITT for the transmission of text, data and still images [2.52].

2.3.1.2 Conversational Teleservices

The *ISDN telephony service* represents a considerable advance in voice communication for both the business subscriber and the private user. Due to the technical characteristics of the ISDN, this new telephony service provides enhanced speech quality and greater convenience: improved signal-to-noise ratio, attenuation unaffected by distance, improved handsfree talking facilities. At a further stage in its development, ISDN telephony could offer a greater voice

Table 2.1. Services in the ISDN with Bit Rates up to 64 kbit/s

Service classes	ISDN services via B-channels (64 kbit/s)	ISDN services via D-channels	Existing services taken over from the telephone network	Services for terminals of dedicated text and data networks
Conversational services	*Bearer services:* ISDN Data Transmission *Teleservices:* ISDN Telephony, Telephone Conference ISDN Teletex ISDN Telefax (Telefax 4) ISDN Teletex Mixed Mode Telewriting Still Image Transfer Service Videotelephony Service[2] ISDN Alarm Services ISDN Teleaction	ISDN Data Transmission (packet-switched) ISDN Alarm Services ISDN Teleaction	Telephony Telefax 2, Telefax 3 Data Transmission with V-interfaces (parallel, serial) Alarm Services Teleaction	For Teletex (2400 bit/s) Possibly for Telefax 4 (9600 bit/s) For Data Transmission with X.21 and X.25 interfaces
Messaging Services	Voice Mail Text Mail Fax Mail			
Retrieval Services	ISDN Videotex		Videotex	Videotex
Distribution Services	Data Distribution Voice Distribution Still Image Distribution			

[2] With highly reduced spatial and temporal resolution

bandwidth, e.g. 7 kHz, and stereo sound – an important service attribute for audio conferencing. The user is also offered a large number of new supplementary services as part of this service (Sect. 2.3.3).

The *ISDN teletex service* is the ISDN version of the already existing teletex service adapted to the ISDN as standardized by CCITT [2.9]. Due to the higher transmission rate of 64 kbit/s, the time taken to transmit a page of ISO A4 or North American legal paper size is reduced to less than 1 s. The protocols and functions for text transmission and presentation (functions of OSI layer 4 to 7) [2.9, 2.14 to 2.18, 2.39 to 2.40, 2.43 to 2.44] are the same as used in the existing teletex service, so that intercommunication between teletex terminals connected to the ISDN and other networks such as circuit-switched or packet-switched data networks is possible. Rapid text transmission with ISDN teletex or a future multimedia document transfer service will become increasingly important for future office communication.

The facsimile service *ISDN telefax (telefax 4)* is used for the transmission and high-quality reproduction of pictures, drawings and handwritten text. As far as the protocols (layers 4 to 7), coding and resolution are concerned, this service will be based on the CCITT Recommendations for telefax 4 service and group 4 facsimile machines [2.8, 2.13, 2.17 to 2.18, 2.22, 2.25, 2.27, 2.33, 2.39 to 2.40, 2.43 to 2.44]. These Recommendations provide for resolutions of up to 300 ppi, optionally 400 ppi up to 1200 ppi; a resolution of 400 ppi corresponds to the quality of present-day office copiers. At this resolution and with black-and-white reproduction, the average amount of pixel-encoded information contained in a page of ISO A4 or North American paper size can be transmitted in about 15 s on ISDN telefax (64 kbit/s). ISDN telefax can thus provide for the first time a practical high-quality facsimile transmission system which can accompany voice communication.

The combined text and facsimile mode of ISDN teletex, *ISDN teletex mixed mode*, is particularly suitable for cost-effective transmission of black-and-white documents consisting of text and graphics or having hand-drawn illustrations, and of letters with letter heading and signature. This service provides two modes of operation, one for the character-coded transmission of texts, and the other for the pixel-coded transmission of graphics and pictorial information (facsimile). The service attributes such as character sets, resolution, communication protocols, transmission time, etc. correspond to those of the ISDN teletex and ISDN telefax services [2.11, 2.13, 2.15, 2.22 to 2.23, 2.28, 2.31, 2.37 to 2.46]. The extension of teletex mixed mode to other information types besides text and facsimile, such as graphics (geometric mode), still images with high resolution and audio (voice annotations), may lead to a multimedia document transfer service in the future.

A further communication service intended primarily for private use is *telewriting*. Short messages are written with an electronic pen on a note pad and sent to the receiver directly or stored in his electronic mailbox. The messages may be reproduced on the domestic television screen by vector graphics using an accessory unit or on other terminals with video screens, e.g. a modified telephone, in the home or office of the addressee. The CCITT has established initial definitions for the use of telewriting as a supplement to telephony [2.64, 2.65]; prototype terminals are undergoing trials in the ISDN.

The *ISDN still image transfer service* is used to transmit individual TV freeze frames. It is also possible to transmit still image sequences using special compression methods so that, depending on content, a new image is reproduced every 1 to 10 s. In this way the impression of moving pictures can be produced to a limited extent ("slow motion").

The *videotelephony service* with 64 kbit/s or 2×64 kbit/s offers users a moving-picture transfer capability for person-to-person, person-to-group and conference communication. However, as the bit rate of 64 kbit/s (or 2×64 kbit/s) is very slow for transmitting moving pictures, the video information to be transmitted must be severely compressed. This means that motion video services of this kind using the video encoding methods currently available [2.66 to 2.68, 2.71] can only provide a picture quality inferior to color television. Consequently, video telephones and associated conferencing equipments incorporate special operating modes for transmitting documents as higher-resolution still images. The possible applications of the 64- and 2×64-kbit/s videotelephony and videoconferencing services require further investigation – also in view of the future broadband full-motion video services planned (see Sect. 2.5). When developing video encoding algorithms, it is important to ensure that intercommunication is possible between different videotelephony services. This can best be achieved, insofar as the encoding methods are incompatible, by switching over the encoding in the terminals (multimode codecs) and if necessary by bit rate adaption in network/service gateways.

The first international standards for videotelephony (service, transfer protocols, terminal) will be published by the CCITT in 1991/92 [2.69 to 2.75]. Prototypes of videotelephones using non-standardized transmission methods are already available. Some service providers, such as Deutsche Bundespost Telekom, are planning to introduce a videotelephony service operating at 128 kbit/s in the ISDN by 1992.

The *alarm services* include facilities for emergency calls. *Teleaction services* are used for telemetering (e.g. meter reading), monitoring and controlling of commercial and domestic installations (heating, electricity, gas, water) and for controlling road traffic systems.

These services generally require transmission paths with high availability, a low bit error ratio, special provision for reliability and data protection and short transmission times.

Teleaction and alarm signals with higher data volumes and higher priority can be transmitted in the ISDN via circuit-switched and via permanent B-Channels. For applications involving very small data streams, transmission in the D-channel is preferable (see Sect. 2.3.2.2).

2.3.1.3 Messaging Services

Voice mail, text mail, fax mail and the message handling service within the videotex service are messaging services incorporating mailbox functions for voice, text, facsimile or data. Using teleservices of this type, information can be sent to communication partners even from unattended terminals by depositing voice, text or graphic information in their personal electronic mailbox. As a supplementary service, the user will be automatically informed by the mailbox system that he has received a message in his mailbox. This notification may be sent to the user over the signaling (D) channel independently of the operational status of the user's terminal and network access or via a free B-channel by a special control message. The user can retrieve individually the messages placed in his mailbox. The electronic mailboxes assigned to individual users or user groups can be implemented in storage facilities connected to the ISDN at central points (cf. Sect. 6.1). Access by the subscriber to the storage centers using the ISDN voice and text/facsimile/data services will be considerably facilitated by the service integration features of the ISDN mentioned in Sect. 1.6.

With suitable equipment, the messaging services can also be used for data transmission during low-charge periods, for multiaddress (broadcast) transmission and for communication between noncompatible terminals (via service interworking unit with conversion and processing functions). Messaging systems of this type are termed "message handling" systems (MHS) by the CCITT and recommendations for these have already been agreed [2.76].

2.3.1.4 Retrieval Services (Teleservices)

ISDN videotex is a further refinement of the present-day videotex service [2.12, 2.63]. It is designed for combined application of the alphamosaic, geometric and photographic modes of operation [2.20, 2.21, 2.47, 2.77] for the efficient transfer of mixed text and graphic information. The processing speed matched to the transmission rate of 64 kbit/s, and the optimized coding methods (reduced redundancy) permit rapid image buildup (within 1 to 20 s depending on the image content, operating mode and coding method) for information retrieval and rapid image input by the information providers at the videotex centers. It is only at a rate of 64 kbit/s that the transmission time of (pixel-coded) images and graphics in the photographic mode is acceptable for the user.

Due to the picture reproduction method of existing domestic television sets, the resolution of the information that can be represented with videotex is today limited to 480 pixels horizontally and 240 pixels vertically. When digital television technology is generally established and picture tubes with higher resolution are used, e.g. for future high-definition television (HDTV), it will be possible to formulate the ISDN videotex standard accordingly for a higher resolution.

2.3.1.5 Distribution Services

64 kbit/s distribution services will not have the same importance as future broadband distribution services (e.g. for television programs). They are only suitable for special applications of voice, data and still image distribution. Due to the unidirectional transmission paths, transmission error correction techniques with error indication in the backward direction and retransmission of information frames (ARQ) cannot be used. The use of such distribution services

is only practicable in conjunction with forward error correction techniques (FEC), or if the messages are transmitted with cyclic repetition and are overwritten in the terminal of the receiver at short intervals. The application of such services is still unresolved.

2.3.2 ISDN Services via the D-Channel

The D-channel of the ISDN is primarily used for signaling between user and network. The D-channel is also used – always taking the signaling priority into account – for transmitting data in packet mode from data transmission services and e.g. alarm or teleaction signals from security and teleaction services.

The available transfer rates of additional services of this kind via the D-channel vary as a function of the signaling load on the D-channel. Average throughputs of up to 10 kbit/s are conceivable on a D-channel with a transmission capacity of 16 kbit/s.

2.3.2.1 Conversational Bearer Services with Low Throughput

To supplement the packet-switched bearer services via B-channels (see Sect. 2.3.1), packet-oriented bearer services with virtual call and permanent virtual circuit facility via the D-channel are subject to CCITT standards [2.53]. These can be used, in compliance with the respective national specifications, for transparent transmission of text and data with low throughput. The above-mentioned restrictions in terms of throughput and priority apply.

2.3.2.2 Conversational Teleservices

Possible teleservices using the D-channel will be *alarm services* for emergency calls and *teleaction services* for remote meter reading etc. and remote control as already described in Sect. 2.3.1.2.

For applications with very small data streams, it is generally preferable in the ISDN to use the signaling channel with its packet-oriented transmission structure. Again, the above-mentioned restrictions in respect of throughput and priority apply.

2.3.3 Supplementary Services

As defined in Sect. 2.2, the *general access attributes* describe ISDN access capabilities relating to the subscriber's access and usable by all services. The *basic service attributes* describe the basic characteristics of a service and are thus fixed components of it. In addition, a user has the option of changing or improving this basic service by means of *supplementary services*. Thus, supplementary services supplement or enhance the basic service attributes.

The supplementary services are subdivided into the following groups in accordance with Fig. 2.5:

- Access related supplementary services,
- Connection related supplementary services,
- Network information related supplementary services.

General access attributes and supplementary services possible in the ISDN are illustrated below using appropriate examples. Table 2.2 provides examples of supplementary services, assigned to five ISDN services. This table is not

Table 2.2. Important Supplementary Services in the ISDN and Their Assignment to Some ISDN Services[1]

ISDN supplementary services	ISDN services				
	ISDN data transmission	ISDN telephony	ISDN teletex	ISDN telefax (Group 4)	ISDN videotex
Access related supplementary services					
Closed user group	×	×	×	×	×
Direct dialing in	×	×	×	×	×
Multiple subscriber number	×	×	×	×	×
Connection related supplementary services					
Call waiting		×			
Registration of incoming calls		×			
Call forwarding unconditional		×			
Call forwarding no reply		×			
Call forwarding busy		×			
Completion of calls to busy subscribers		×			
(International) Freephone service	×	×	×	×	×
Reverse charging	×	×	×	×	×
Conference calling		×			
Abbreviated dialing	○	○	○	○	○
Call hold service		×			
Outgoing call barring		○			
Incoming call barring	×	×	×	×	
Redialing		○	○	○	○
Multi-address calling	○		○	○	
ISDN networking services (City-wide centrex)	×	×	×	×	×
Network information related supplementary services					
Calling line identification presentation		×	× ○	× ○	
Date and time			×	×	○
Malicious call identification		×			
Advice of charge	×	×	×	×	× ○
Announcements		× ○			○

× Implementation in switching equipment;
○ Implementation in terminals or special service centers.

[1] These are only examples of possible ISDN services and supplementary services. A particular ISDN may not offer all these ISDN services and supplementary services. Some telecommunication carriers may offer ISDN-based services which have not been defined or identified by CCITT.

exhaustive. It also indicates whether the individual supplementary services should preferably be implemented in switching equipment, in terminals or in special service centers (e.g. videotex centers).

General access attributes of the ISDN

- Multiservice operation: simultaneous use of several services. In this case several connections with the same or different services to one or more destinations are in operation at the same time.
- Change of service during the connection, e.g. change from voice communication to facsimile transmission.
- Switched connection.
- Semipermanent connection.
- Permanent connection.

Supplementary services

Access related supplementary services
In contrast to the general access attributes of the ISDN, access related supplementary services are provided only in connection with services for which they are applicable and activated by the subscriber.

- *Closed user group*: the users form groups with specific, agreed restrictions for access to and from users of the public network (private branch exchange functions) [2.78].
- *Direct dialing in* to extensions in private branch exchanges [2.79].
- *Multiple subscriber number* for selecting specific terminals on the passive bus of the called user, if necessary with distinctive ringing [2.80].

Connection related supplementary services
These are ISDN supplementary services for *quick and convenient connection set-up, special completion of connection, information transfer with information storage, restriction of connection (call barring)*, and *special connections* and *conversions (service, network and terminal adaptations)*.

- Quick and convenient connection set-up and convenient communication
  - *Abbreviated dialing*: for frequently used directory numbers the user can employ two-digit abbreviated numbers for connection set-up.
  - *Redialing*: the directory number last dialed is stored. Dialing is repeated by pressing a specific key or automatically after a preset time.
  - *Fixed destination call*: after actuating any key or merely by lifting the handset, a connection is set up automatically to a line with a specific directory number.
  - *Handsfree dialing*.
  - *Handsfree speaking* via a separate microphone.

- Special completion of connection
  - *Completion of calls to busy subscribers*: the calling user can activate this feature if a called line is busy. When the called line is released the network will inform the calling user that the called party is no longer busy. On request of the calling user a recall to the called party will be set up by the network [2.81].
  - *Call waiting*: during an existing connection the called user is audibly and/or visually informed of the presence of a further request for connection, possibly with indication of the calling number. He can accept the second connection within a distinct time interval [2.82].
  - *Call forwarding unconditional*: the user can forward all incoming calls to any other line by entering its directory number [2.83].
  - *Call forwarding no reply*: the user can enter any other directory number, to which an incoming call is forwarded if he does not accept this call within a certain time (e.g. three rings) [2.84].
  - *Call forwarding busy*: incoming calls are forwarded to the entered directory number if the called subscriber is busy [2.85].
  - *Selective call forwarding*: calls from certain directory numbers specified to the switching system by the subscriber are not forwarded.
- Information transfer with information storage
  - *Absent subscriber service* with or without storage of the replies: the functions are the same as those of automatic answering equipment.
  - *Registration of incoming calls*: if calls are made while the called subscriber is absent, the date, time, and directory number of the calling subscriber are recorded for the information of the called subscriber.
- Restriction of connection (call barring)
  - *Total call barring*: barring of the line for all outgoing and incoming connections on request by the subscriber.
  - *Outgoing call barring*: on request by the subscriber for e.g. intercontinental connections, international connections or long-distance connections [2.86].
  - *Incoming call barring*: on request by the subscriber, barring at certain times (e.g. do-not-disturb service) or permanent barring (line for outgoing calls only) of all connections or services specific connections.
- Special connections
  - *International freephone service, Freephone service*: the called subscriber accepts all charges for incoming calls. Certain directory numbers are made available for this, such as *Service 800* in the USA [2.87].
  - *Reverse charging* (case by case): the called subscriber can accept the connection charges on a case-by-case basis during the call or before accepting the call [2.88].
  - *Wakening call service*: the user arranges for a wakening call on his line by specifying the date and time.
  - *Multi-address calling*: for transferring one-way messages, several main station lines are called simultaneously or consecutively (information distribution).

- *Call hold service*: the user can switch the existing connection with a second user to a wait status, in order to set up an additional connection to a third user for inquiry purposes or to accept a waiting connection [2.89].
- *Conference calling*: connections with three users or more at the same time, each being able to communicate with all the others [2.90].
- *ISDN networking services* (*city-wide centrex*): PABX-related supplementary services to support easy set-up and control of the call (e.g. *private numbering plan* [2.91]), to increase the accessibility of the communication partner, for access authorization, special accounting, directory inquiries and other facilities provided to subscribers by means of specially equipped public exchanges.
- Conversions: If no conversion function is provided, interworking is only possible between compatible services, networks and terminals. To support unrestricted (open) communication, conversion functions have to be provided for service, network and terminal adaptation: conversion of bit rates, protocols, interfaces, etc.

Network information related supplementary services
These include possible ISDN supplementary services such as *charge information, network information, directory inquiries*, and *identification*.

- Charge information
  - *Advice* of the current or accumulated *charge* of a call as charge units or amount of money during a connection or after disconnection [2.92].
  - *Printed record of call charge* (in text communication services) after disconnection.
  - *Toll ticketing*: itemized billing for each connection and service (date, time, directory number of the called subscriber).
- Network information
  - *Announcements*: indication or announcement e.g. of a changed directory number.
  - *Call progress signals* for user guidance during connection set-up.
  - Indication of *date and time* on connection set-up.
  - *Incoming message waiting indication*: the user is automatically informed by the mailbox system that a message has been deposited in his mailbox.
- Directory inquiries
  - *Public recorded information services* such as weather, sports results, news.
  - *Telephone directory inquiry services*.
  - *Directory inquiry services* for telex, teletex, other services.
- Identification
  - *Malicious call identification*: recording the directory number of malicious callers.
  - *Calling line identification presentation* at the called subscriber's terminal [2.93].
  - *Connected line identification presentation* (for the calling user) [2.94].

2.3.4 Existing Services from the Telephone Network in the ISDN

In the ISDN, the services of the telephone network can be retained alongside newly introduced services at least for a transitional period, so that existing terminals of present-day services of the analog telephone network can continue to be used.

Consequently, the existing telephone service with its analog telephones can also be initially accommodated in the ISDN. The subscriber of this service is assigned an analog line at his ISDN exchange (Fig. 2.6, see also Sect. 6.1). All text and data services of the existing telephone network can be handled as before on this line. This applies to data transmission services with modems (in accordance with CCITT V.-series Recommendations, some of which correspond to EIA standards, e.g. RS-232-C) at the transmission rates currently defined for the telephone network, and also to the telefax and videotex services.

However, these services can also be provided at the ISDN access via a special terminal adaptor (see Sect. 4.1.4) with t/r ("tip and ring") interface

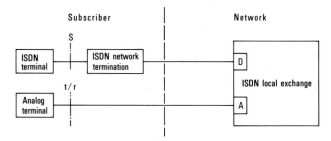

Fig. 2.6. Digital and Analog Subscriber Lines Connected to an ISDN Local Exchange. A analog line module, D digital line module, t/r analog user-network interface, S ISDN user-network interface

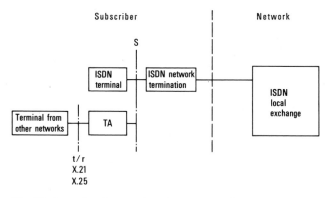

Fig. 2.7. Examples of Adaptation of Terminals from Other Networks to the ISDN Subscriber Line. S ISDN user-network interface, t/r analog user-network interface, X user-network interface as specified for data networks, TA terminal adaptor

(Fig. 2.7). In these cases the analog modem signals are digitized in the adaptor unit and are switched and transmitted as digital signals. The service and terminals remain the same, but using the service via the ISDN access has the advantage that ISDN capabilities can also be used for the existing service, e.g. the integrated multichannel access or the improved call establishment by means of an ISDN terminal within the user station (cf. Sect. 1.6). For example, due to the multichannel capability of the ISDN line, the telephone circuit is not blocked while videotex is being used.

2.3.5 Services in Conjunction with Terminals from Dedicated Text and Data Networks

As far as required, the ISDN will also provide access possibilities, via terminal adaptors, for existing terminals intended for operation in dedicated switched text and data networks (Fig. 2.7). Here, too, the user who needs ISDN access to make use of new ISDN services can enjoy the advantages of the integrated ISDN subscriber line even if he retains some conventional terminal equipment. The main options for access to the ISDN are as follows:

- Data terminals with interface in accordance with CCITT Recommendation X.21 for circuit-switched traffic (cf. Table 2.3).
- Data terminals with interface per CCITT Recommendation X.25 for packet-switched operation (cf. Table 2.3).
- Teletex terminal equipment (2.4 kbit/s).
- Telefax 4 terminal equipment.

Although no changes will be made to these terminals, the subscriber who connects them to the ISDN will receive a service that differs from that provided in the text and data network. This is due to the different characteristics of the network; for example, the connection set-up times may be different.

Table 2.3. User Classes in Circuit and Packet Switched Data Networks[b]

Terminals with interface as per CCITT Rec. X.21		Terminals with interface as per CCITT Rec. X.25	
User class[a]	Speed in bit/s	User class[a]	Speed in bit/s
4	2400	8	2400
5	4800	9	4800
6	9600	10	9600
7	48000	11	48000
19	64000	13	64000

[a] Numbering as per CCITT Rec. X.1 [2.95].
[b] Most important for communication via ISDN.

2.4 Service and Network Interworking Facilities

In specifying services, the aim is to facilitate communication between as many users of a network as possible. The same applies to intercommunication between terminals of different services and between terminals connected to different networks. This is achieved by means of service and network interworking facilities as required (Fig. 2.8).

2.4.1 Intercommunication Within the ISDN

It goes without saying that voice communication between existing analog telephones and ISDN telephones should be possible for as long as analog telephones are in use. When connected to subscribers of the existing public telephone network, the ISDN subscriber naturally cannot fully utilize the supplementary services of the ISDN telephony service. For example, *calling line identification presentation* (cf. Sect. 2.3.3) cannot be realized.

In connection with services for non-voice communication, the following intercommunication relations are important within the ISDN:

- Within the ISDN teletex service: teletex terminals (64 kbit/s) – teletex terminals (2.4 kbit/s) with terminal adaptor.
- ISDN teletex mixed mode – ISDN teletex basic mode, ISDN telefax (telefax 4), and future multimedia document services.
- ISDN telefax (telefax 4) terminals – telefax 3 terminals with terminal adaptor.

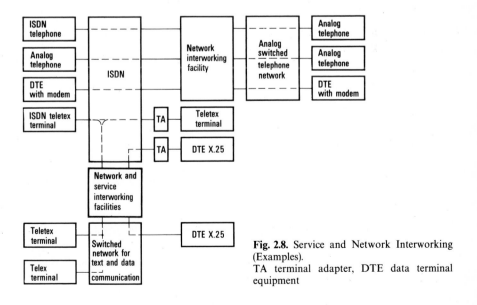

Fig. 2.8. Service and Network Interworking (Examples).
TA terminal adapter, DTE data terminal equipment

2.4.2 Intercommunication Using Network Interworking Facilities

Just as within the ISDN voice communication between telephones connected to an analog line and to an ISDN line must be possible, so telephones connected to the analog telephone network or to not yet converted analog parts of the network must be able to communicate with telephones connected to the ISDN. For the ISDN subscriber, essentially the same restrictions as mentioned in Sect. 2.4.1 apply.

In countries with dedicated text and data networks, interworking facilities between the ISDN and these networks are of considerable importance. For example, they enable a computer connected to the data network to be accessed from within the ISDN. All the teletex terminals of both networks can communicate with each other, and there exists (via teletex) a connection from the ISDN to the telex service. Also, intercommunication of all telefax terminals connected to ISDN and different networks is supported.

2.5 Services with Higher Bit Rates

Besides the services based on the B-channels (64 kbit/s) or on the D-channel of the ISDN subscriber line, as the ISDN develops further to the broadband ISDN services with higher bit rates can be expected.

The CCITT has already defined ISDN bearer services with the so-called H-channel bit rates [2.52] (see Sect. 4.2): H0 = 384 kbit/s and H11 = 1536 kbit/s for digital systems in USA, Canada, Japan, and H12 = 1920 kbit/s for digital systems in Europe. Teleservices using those bit rates are under discussion in the CCITT. Examples of applications of those services are broadcast sound transmission, video conferencing (connection of conference studios), and data transmission (e.g. file transfer, electronic newspaper).

Virtual channels with bit rates up to about 130 Mbit/s via a 155-Mbit/s interface and with even higher bit rates via a 600-bit/s interface will be provided by broadband ISDN using asynchronous transfer mode ATM (cf. Sect. 4.5). Channels of this kind offer a wide range of new services [2.96 to 2.104].

The most important broadband ISDN services and applications for business communications from the point of view of today's user are:

– broadband connection-oriented bearer services for high-speed data communication including the connection of local area networks (LANs) and metropolitan area networks (MANs),
– broadband connectionless bearer services (with switching of individual data packets) for setting up MANs using public network transmission and switching facilities,
– broadband facsimile as a precursor of a future multimedia document transfer service (also as color fax once standardization has been agreed),

– broadband videotex for retrieval of multimedia information incorporating high-resolution still images, voice and moving pictures (film scenes) in addition to text, data and graphics,
– broadband videoconferencing (studio and workplace conferences).

For the domestic sector, the most important broadband services are:

– broadband videotelephony with picture quality comparable or superior to present-day TV,
– broadband distribution services for high-definition TV (HDTV) and in some cases broadband TV distribution services (supplementing existing cable TV networks).

3 ISDN Structure

3.1 Network Organization

Communication networks essentially comprise three components (Fig. 3.1): the subscriber line network, the exchanges and the trunk network.

- The *subscriber line network* connects the subscribers' terminals to the local exchanges to which they are assigned. The subscriber lines may be provided on a per subscriber basis, as in most of the telephone network, or shared by several subscribers by means of multiplexing. The latter approach is frequently adopted in the subscriber line area of text and data networks, as the lower density of these networks results in greater distances between subscriber and exchange. Copper wire pairs 0.4 to 0.8 mm in diameter are used for the subscriber lines (see Sect. 7.4.3). In the subscriber line network of the telephone system, each subscriber is served by a two-wire circuit. In text and data networks, due to the greater distances involved, data transmission systems using modulated carriers or baseband transmission are employed. These systems exist in both two-wire and four-wire versions. Four-wire circuits are also provided in some private networks.

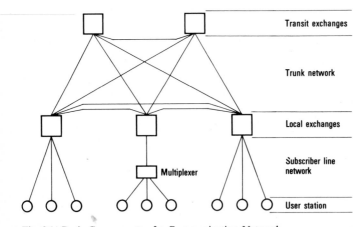

Fig. 3.1. Basic Components of a Communication Network

- The *exchanges* establish the connection, depending on the directory number dialed, between the subscriber line and an outgoing trunk circuit (local exchange, local office) or interconnect the trunk circuits of different trunk groups (transit or tandem exchange). In many cases the same exchange carries out both functions (combined local/transit exchange). As it is not economic in national networks for all the local exchanges to be interconnected in a mesh network, local networks are formed within defined geographical areas. The long-distance traffic of these local networks is fed in concentrated form to the long-distance (toll) network which interconnects the regional centers. This network arrangement creates a distinction between exchanges for local and long-distance traffic.
- The *trunk network* interconnects the exchanges. *Local trunks* connect the exchanges within the local network, *long-distance* (*intertoll*) *trunks* the exchanges of the toll network. The local and toll networks are linked via *trunk junctions* (*called toll connecting trunks* in the USA).

In addition, functions and equipment for control and maintenance are required for operating a network. These include call charge registration, call data administration, quality and grade of service measurement, and maintenance of the network equipment.

To support these tasks associated with operating a communication network, the initial recommendations for a Telecommunications Management Network (TMN) are being prepared by the CCITT in the study period 1989–1992 [3.1]. The TMN concept basically envisages a self-contained system via which the information from the various network components is collected and processed and via which the components can be configured and controlled by the operator and also by the user.

The "intelligent network" concept has been developed with the aim of providing greater flexibility for introducing new services and modifying existing ones [3.2]. The functions required for controlling a service – such as analyzing a service requirement, selecting the necessary network components and controlling the logical linking of these components – are not incorporated in the individual exchanges in the network, but in separate service control points (cf. Sect. 6.3.6). These centers, to which the users also have access, are connected to the exchanges via Signaling System No. 7 (see Sect. 6.3.1). The logic for controlling a service must be implemented in the form of rigorously defined functional components describing switching sub-sequences, thereby also allowing the network users to modify a service independently.

3.2 Initial Situation for the ISDN

The ISDN cannot be regarded independently of the existing networks. These are the *telephone network*, designed primarily for handling telephone traffic, and the special networks for text and data traffic.

3.2.1 Telephone Network

Telephone networks are mostly configured as combined star/mesh networks and are subdivided into several hierarchical levels (Fig. 3.2). The number of hierarchical levels depends on a number of factors, important variables being the size of a country and the subscriber density.

Each exchange is connected to the next exchange above it in the hierarchy via the trunk groups of the star network (heavier lines in Fig. 3.2). In addition to these circuits forming the so-called final trunk group, there are high usage trunk groups which connect exchange at the same and different hierarchical levels via the mesh network. These high usage trunk groups are provided between two exchanges if there is sufficient traffic volume to justify them on economic grounds. Whether such high usage trunks are provided also depends on the technical capabilities of the switching systems. While modern SPC-controlled exchanges generally impose no limitations, the older electromechanical systems will not allow every economically viable high usage trunk group to be implemented. This may be because the maximum number of connectable trunk groups is limited or there are restrictions as regards call number interpretation.

Telephone networks are further subdivided into local and long-distance (toll) networks. The local network is used for handling traffic within a defined geographical area, e.g. a city, and may itself be subdivided into several hierarchical levels. Long-distance and international traffic is handled via the toll network. This division into local and toll network is attributable to the different technical requirements of the two parts of the network. Whereas with analog transmission and switching a two-wire arrangement is usually adequate in the local network due to the shorter distances to be spanned in the toll network a four-wire

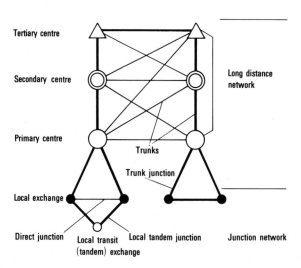

Fig. 3.2. A Four-Level Hierarchical Network Example (Mainly CCITT Terms)

arrangement is necessary. Furthermore, lower requirements are placed on the local exchanges than on the long-distance exchanges in terms of traffic routing and call charge registration. However, the progressive digitization of the networks, with four-wire transmission and switching throughout, and stored-program control of the exchanges make it possible to transfer toll network functions to the local exchanges.

By connecting voice-band data modems to the subscriber line, the telephone network can also be used for data transmission. For this purpose the digital data signals must be converted by the modem to voice frequency signals [3.3].

However, there are a number of limitations affecting data transmission over the telephone network, which for certain applications have no satisfactory solution. These include:

- the limited data transmission rate (mainly up to 9600 bit/s for modems currently in operation, and up to 19 200 bit/s in the future),
- the relatively high bit error ratio due to transmitting data as analog signals,
- connection setup times (including entry of the call number) in excess of 10 s,
- possible network blocking during peak traffic periods.

Nevertheless, in many countries data transmission over the telephone network is one of the most frequently used forms of data communication. This is due among other things to the fact that the equipment already installed for the telephone service can be used instead of setting up a separate network.

The *videotex service* [3.4] and the *telefax service* [3.5] for facsimile transmission are two other services mainly handled via the telephone network. For videotex, the telephone network establishes the connection between the videotex users and the information centers; for the telefax service, the connections are dialed-up as for a telephone call and then used for facsimile transmission.

In all these services the telephone network is used solely for information transport and does not provide any supplementary services such as *closed user group* (see Sect. 2.3.3) in addition to the service attributes for the telephone system.

3.2.2 Text and Data Networks

The special requirements of text and data communication, such as higher data transmission rates, shorter connection set-up times and lower error ratios for data transmission as compared with the telephone network, have led many countries to construct separate networks for text and data communication.

These networks employ different through-connection techniques (circuit switching, packet switching) and, due to the smaller number of subscribers, have fewer hierarchical levels than the telephone network. Owing to the lower subscriber density, it is usually only viable to construct exchanges in large population centers. More remote subscribers therefore have to be connected via comparatively long subscriber lines. To save costs in this part of the network, concentrators and multiplexers are generally employed; these concentrate the traffic on the subscriber lines and forward it to the exchanges over a reduced number of lines.

For the text and data networks the CCITT has produced separate Recommendations [3.6] which take into account the special requirements of these networks. For example, these networks provide a number of supplementary services, such as *closed user group, reverse charging*, and *multi-address dialing*, and permit synchronous and asynchronous operation of the connected terminal equipment as well as half-duplex and duplex operation.

The main advantages of networks operating on the *packet-switching* principle include matching of the different transmission procedures and rates of the connected terminals, short connection set-up times, use of the subscriber lines for several connections at the same time, and call charge advantages in certain applications, e.g. in the interactive mode.

In many data transmission applications, the connection set-up times and the risk of the desired connection not being immediately available due to lack of free transmission paths or due to the subscriber line being busy, are unacceptable. Moreover in many cases the advantage provided by a switched network of being able to reach a large number of lines is irrelevant because the actual traffic relations are limited to a few specific communication partners. For these applications, dedicated circuits become a viable option. These circuits are used to interconnect two data terminal equipments on a permanent basis. Connection set-up times are virtually negligible, the transmission path is always available and the line cannot be occupied by other calls. Some links of this kind are routed via analog transmission paths; as with data transmission over the telephone network, these require modems at the terminals to convert the digital data signals to voice frequency signals.

3.3 Functions of the ISDN Network Components

The ISDN is based on the digital network components of the telephone network (see Sect. 3.8). These facilities, designed to meet the requirements of the telephone network, have to be supplemented in order to realize the potential of the ISDN.

3.3.1 Subscriber Lines in the ISDN

In the ISDN all information is transmitted digitally, even on the subscriber line (Fig. 3.3a). (This represents a considerable departure from the digital telephone network, in which information is transmitted in analog form as far as the local exchange; see Fig. 3.3b). The existing copper wires will be used as transmission paths. With suitable transmission methods (see Sect. 7.4.3), the copper wire pair of the subscriber line can be used to transmit the net bit rate of 144 kbit/s

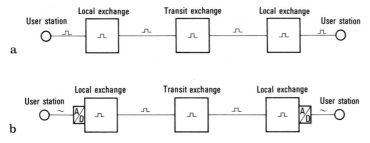

Fig. 3.3 a, b. Network Concepts. **a** ISDN; **b** Integrated digital network (IDN).
~ Analog signal, ⎍ Digital signal, A/D Analog/digital converter

required for the ISDN basic access. In the subscriber line network, no additional expenditure, e.g. for regenerative repeaters, is normally required.

3.3.2 Local Exchanges in the ISDN

In the local exchange, the enhanced communication facilities provided by the ISDN result in a more complex subscriber access than that of the digital telephone network and place additional requirements on call control in the establishment and clearing phases (see Sect. 4.3.5). The essential functions of the subscriber line circuit in the exchange include splitting up the 144 kbit/s information stream into the two 64 kbit/s channels and the 16 kbit/s channel and separate processing of the two 64 kbit/s channels. These are processed separately because they can be switched through to different destinations and can be used for different services. It must be possible, during an established connection, to add-on a second 64 kbit/s channel and to change the service within a channel. This means that subscriber signaling must also be possible during an established connection. The ISDN differs in this respect from the telephone network, in which only call charge information and the criterion for clearing a connection are transmitted in the course of a call. The ISDN also enables several terminals to be connected to a subscriber line under a single directory number. For incoming calls, this necessitates an exchange of information between local exchange and terminal prior to through-connection, and sometimes between the terminals at either end in order to ascertain the terminal to which the incoming call can be assigned.

3.3.3 Signaling Between ISDN Exchanges

Signaling between ISDN exchanges requires a more extensive range of features compared with conventional telephone signaling. The method used is CCITT Signaling System No. 7 with central signaling channels between the exchanges, supplemented by additional functions for ISDN (see Sect. 6.3). The formats for identifying the services have been expanded, and changes and additions have been made to the signaling protocols for controlling new service attributes. Some of the signaling information to be passed between exchanges only relates to the originating and destination exchanges. To handle this information, end-to-end signaling has been introduced as a new function. End-to-end messages are identified by a special code and are merely passed on by the intervening transit exchanges.

3.3.4 Operation and Maintenance in the ISDN

The operation and maintenance functions for the ISDN network components are largely subject to the same requirements as those affecting the digital telephone network, particularly where the exchanges and the trunk network are concerned. The main functions are detailed in Sect. 6.2.7. Additional requirements basically concern the subscriber line network, as the analog subscriber access is now replaced by a digital one. The subscriber access consists of several functional groups (see Sect. 4.1), such as the line terminal equipment at the exchange, the network termination unit on the subscriber side and the subscriber line. In order to enable a defective network component to be identified in the event of a fault, the subscriber line circuit is divided into several test sections within which test loops can be inserted [3.7]. These test loops can be remotely controlled from the exchange. By comparing the transmitted and received data in the test loops it is possible to determine whether a fault is present in the test section in question.

3.4 Network Dimensioning

3.4.1 Basic Considerations

The aim of network dimensioning is to design the network equipment used jointly by the subscribers, i.e. the exchanges and the trunk network, in such a way that the desired connections can be established without significant blocking even at peak traffic periods [3.8]. In the case of the exchanges, it is necessary to specify the number, position, size and traffic processing capacity; for the interexchange trunks, it is necessary to determine the number of circuits combined to form a trunk group.

In dimensioning switching systems a distinction is drawn between systems operating in the delay mode and those operating in the loss mode [3.9], depending on how switching tasks are handled when blocking occurs.

In the *delay mode*, if blocking occurs the subscriber can wait for a connection path or other switching facilities to become free. Information concerning the mean length of the delay, the probabilities of delays occurring and of exceeding specific delay times is used to describe the grade of service. Exchange control equipment normally operates in the delay mode.

In the *loss mode*, seizure requests which encounter blocking are rejected. The subscriber receives the busy signal. In this case the percentage loss – the ratio of the number of rejected calls to the total number of calls expressed as a percentage – indicates the grade of service. Switching networks employing circuit switching normally operate in the loss mode. If no free connection path can be found within the switching system ("internal blocking") or if no free serving trunk is available ("external blocking"), the seizure request is rejected (Fig. 3.4).

For network dimensioning it is therefore important to specify a suitable grade of service, i.e. the acceptable loss in the case of circuit switching. For trunk groups with full availability, the CCITT recommends a loss of 1%, for which the traffic intensity in erlangs taken as the basis for dimensioning must be calculated from the peak traffic period values of the thirty heaviest days of the year [3.8, 3.10]. This ensures that the loss on most days of the year is below the 1% value and that there is generally sufficient spare capacity available for high-load days. This is important, as "busy" conditions frequently occasion repeated seizure attempts, resulting in additional loading on the control facilities and generating extra dummy traffic on the pre-assigned connection paths in the network [3.11].

In addition to the dimensioning-related losses within a network, losses attributable to the subscriber also occur, such as those caused by "called subscriber busy" and "called subscriber does not answer" conditions (Fig. 3.4). These losses can be reduced in the ISDN by introducing supplementary service

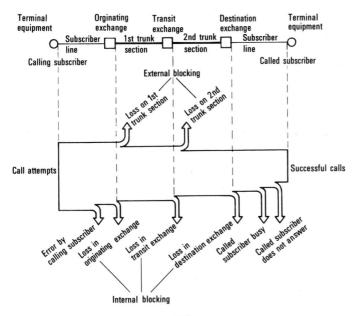

Fig. 3.4. Losses When Setting up a Call

attributes, such as "automatic callback on busy", and voice message handling systems (see Sect. 3.4.2.1).

3.4.2 Effects of Service Integration

3.4.2.1 Holding Time, Traffic Intensity, Busy Hour Call Attempts

In the ISDN voice calls will predominate. The mean holding time of such calls[1] (Fig. 3.5) is substantially determined by subscriber behavior; in contrast to data or text traffic, different bit rates for speech coding (see Sect. 7.2.1) do not alter the mean holding time. Only the shorter call establishment times in the ISDN (1 to 2 s compared with up to about 15 s in the analog telephone network) bring about a reduction in the holding time. The typical figure of approximately 100 s for the mean holding time of voice calls will thus remain substantially unchanged in the ISDN. The same considerations apply to calls with retrieval services such as videotex, and certain types of interactive calls using computers. Here it is mostly the subscriber response times which determine the mean holding time; the duration of information input and output, though affected by the transmission speed (bit rate), represents only an insignificant proportion of the overall holding time of such calls.

The effect of transmission rate on the holding time is much greater for calls which do not depend on subscriber response times but in which it is only the quantity of information and the transmission rate which determine the mean holding time. This applies, for example, to calls for text, data and facsimile transmission. The mean holding times of these calls, for instance typically 10 s for 2.4 kbit/s teletex, – already short compared with voice traffic – will be reduced still further with the transition to the uniform transmission rate of 64 kbit/s (Table 3.1).

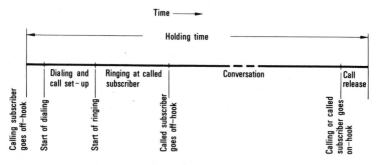

Fig. 3.5. Composition of the Holding Time for a Successful Voice Call

---

[1] Call is any attempt to seize a switching equipment or trunk, irrespective of the success of this seizure attempt.

Table 3.1. Estimated Traffic Volumes in the Busy Hour for Different Communication Services with a Transmission Bit Rate of 64 kbit/s

Service	Mean holding time s	Mean BHCA per line, outgoing and incoming calls per hour	Mean traffic intensity per line, outgoing and incoming Erlangs
Videotex	300	0.36	0.03
Telephony	100	3.6	0.1
Data transmission	15	24	0.1
Facsimile	10	7.2	0.02
Teletex	2.5	7.2	0.005

BHCA Busy hour call attempts

Less pronounced are the effects of service integration on the total traffic in the network. Due to their mainly short holding times, text and data calls generate little traffic compared with voice traffic for the same frequency of calls. For example, for the same number of connections the traffic intensity of 64 kbit/s teletex connections is only about one fortieth of the traffic intensity of speech connections.

Of greater significance, on the other hand, are the effects of service integration on the number of busy hour call attempts (BHCA). Especially for interactive calls in data traffic, call frequencies per subscriber line many times greater than the values hitherto encountered in the telephone service can be expected (Table 3.1). This is due to the fact that these applications are mainly prevalent in the business sector where intensive utilization of the existing terminals is an economic necessity. Furthermore, a single call only lasts for a few seconds, and so considerably more calls can be set up within a given time than in telephony.

Generally speaking, the traffic requirements will change in the following way compared with the telephone network [3.12]:

- The traffic volume to be processed per line will increase due to the text and data services.
- The mean holding time will be reduced.
- In the network the number of BHCA will rise considerably more than the traffic carried. The switching equipment in the network will have to provide enhanced control performance.

Further effects will result from the following:

- Implementation of *supplementary services* (see Sect. 2.3.3), such as completion of calls to busy subscriber, call forwarding/diversion and call waiting.
- Implementation of *mailbox services* with storage of speech and text information which is selectively retrieved on request (see Sect. 2.3.1.3).
- The possibility of change of service during a call, or of the choice of an alternative service in the case of subscriber busy (see Sect. 2.3.1).

Although these facilities initiate additional control procedures in the network,

Table 3.2. Call Mix in the Public Telephone Network

Distribution of calls	in %
Error by the calling subscriber, e.g. off-hook without dialing, incomplete dialing	13
Congestion in the trunk network	6
Called subscriber busy	18
Called subscriber does not answer	8
Successful calls	55

they also increase the availability of the desired communication partner and thus reduce the number of unsuccessful call attempts. They will have the effect of changing the existing pattern of telephone calls [3.13] (Table 3.2) to provide a higher proportion of successful calls.

3.4.2.2 Shared Use of the Network Equipment

A further traffic-related advantage of service integration arises from the sharing of the network's switching and transmission equipment by the different services. This applies in particular to subscriber lines which represent a significant proportion of the capital costs of a network. In the ISDN the copper wire pair of the subscriber line can handle two services simultaneously (e.g. telephony and videotex). Although it is even now possible for more than one service to be handled in the same transmission system, separate channels are provided for these services within that system. Full sharing is only possible if each channel of the system is accessible to each service. This confers the following advantages:

- Text and data service traffic, low in volume compared with the telephone service, can also use the economical high usage interexchange trunk groups installed for bulk telephone traffic. A single service does not in itself always generate a sufficient volume of traffic to justify a high usage trunk group. This concentrating of traffic becomes increasingly important when digital transmission systems are used, because they provide a high channel capacity (30, 120, 480, 1920 or 24, 96, 672 channels or multiples thereof, each operating at 64 kbit/s, see Sect. 7.2, Fig. 7.2).
- Different busy hour times and other traffic fluctuations in the individual services largely balance each other out if common trunk groups are used for all services. The network which only carries light telephone traffic during the night, can also be used for transmitting text and data.

3.4.3 Traffic Routing

There are usually several possible routes within a network between the origin and destination of a call. If the traffic volume between exchanges is sufficiently great, high usage trunk groups are set up which handle 80 to 90% of the offered

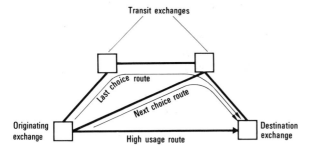

Fig. 3.6. Traffic Routing Schematic

traffic. For the remaining 10 to 20% of the traffic, next choice routes and last choice routes via one or more transit exchanges are available (Fig. 3.6). The aim of traffic routing is to find the most suitable path for a call from these possible alternative paths and seize it, provided at least one circuit is still free in the desired trunk group. When setting up calls between ISDN subscribers who require a wholly digital link, care must be taken with traffic routing in mixed analog/digital networks to ensure that only digital paths are selected.

In both local and long-distance network, digitalization has the following effects on traffic routing:

- The channel capacity of the basic digital system (30 traffic channels in the 2 Mbit/s system or 24 in the 1.544 Mbit/s system; see Sect. 7.2.3) is high compared with both analog local trunks and an FDM system with e.g. 12 voice channels, and raises the economic threshold for installing high usage trunk groups. The traffic previously handled on small high usage trunk groups (less than 10 circuits) becomes transit traffic.
- For transmission engineering reasons, four-wire switching is anyway necessary for long-haul traffic, even with analog switching. As digital switching systems basically operate on a four-wire basis, the additional costs for implementing the transit function are reduced compared with analog technology in which four-wire switching itself means increased expenditure. Instead of separate transit exchanges, transit points can be installed throughout the network on the site of digital exchanges, thereby considerably increasing the options for alternative routing.

Overall, transit traffic in the network will increase somewhat; this is offset by the fact that more traffic can be handled at lower levels of the network [3.14].

An additional refinement of traffic routing results from the increased efficiency of Signaling System No. 7 used in the ISDN. It provides among other things information concerning the originating address of a call and the service indicator and thus enables more effective traffic routing, particularly under overload conditions (see Sect. 6.3). For example, traffic to overloaded destination exchanges can be limited in the originating exchange, or specific services can be prioritized in the event of overload.

Traffic routing methods can also be used in which the routing instructions are modified to suit the time of day or the load situation of the network (dynamic or adaptive traffic routing). The two methods can be combined. In comparison with rigid traffic routing, the dynamic and adaptive methods result in better utilization of network resources and consequently provide savings in the network [3.15].

However, even with a high degree of integration of services, not all parts of the network will provide the same range of capabilities, because not all the network nodes will provide the facilities required for the less frequently used services (see Sect. 6.2.9). Traffic routing must take this into account. It may come about that for some services fewer high usage trunk group options are available for traffic routing than for telephony.

3.5 Interworking with Other Public Networks

In addition the telephone network, many countries have separate networks for text and data communication (see Sect. 3.2.2).

Special services will still be handled via these networks for some time to come. Even when the ISDN is widely established, several of these networks and large parts of the analog telephone network will continue to exist.

In order to provide general availability of communication partners, it is therefore desirable for ISDN subscribers to be able to interwork with subscribers using corresponding services in existing networks. This presupposes conversion facilities (gateways) between the existing networks and the ISDN (see Sect. 6.2.9).

3.5.1 ISDN and the Analog Telephone Network

Due to the number of subscribers involved and hence also the frequency of traffic relations, interworking between the ISDN and the telephone network is of paramount importance. The ISDN is based on the digital telephone network; it is embedded in this network and uses its network components. The facilities for conversion between the analog and digital telephone networks are also available between the ISDN and the analog telephone network (Fig. 3.7). As the numbering plan of the telephone network is also used for the ISDN there are no fundamental numbering problems, and no access codes are required for crossing the interface between the ISDN and the analog telephone network. It is merely necessary that no analog network equipment be used between two ISDN subscribers who require a transparent connection. The calling subscriber must indicate this prior to call establishment (see Sect. 4.3.3.2).

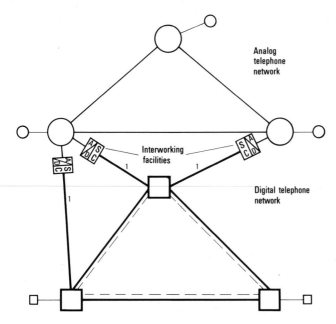

Fig. 3.7. Interworking Between the Analog and Digital Telephone Networks.
[1] Voice circuit signaling, adapted to suit the analog exchanges

Analog   Digital
  o   □   User station
  ◯   □   Exchange
──────── ───────   Trunk circuit

A/D Analog/digital converter, SC Signaling converter, – – – Signaling channel for CCITT Signaling System No. 7

3.5.2 ISDN and Public Data Networks

In contrast, conversion to the public circuit-switched or packet-switched data networks involves interworking with networks having their own numbering systems, their own signaling methods and protocols, and transmission rates which differ from those of the ISDN.

Interworking between the ISDN and these networks therefore necessitates matching of the signaling protocols and the transmission speeds (e.g. conversion from 64 kbit/s to 2.4 kbit/s text transmission). The type of internetwork gateway is affected among other things by the extent to which existing public data networks and their services are integrated into the ISDN.

The possibilities for interworking with these networks are described in Sect. 4.4.1.

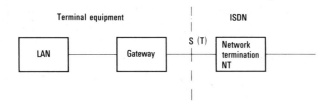

Fig. 3.8. Connection Between Local Area Network and ISDN

3.6 Interworking with Private Networks

Analogously to the public network situation, PBX manufacturers offer private ISDNs for the area of private use by companies and authorities. It is also possible to connect business network subscribers directly to a subscriber exchange in the public network (Centrex; see Sect. 6.4.1). Due to the many and varied communication requirements placed on private ISDNs, a variety of network components are available for their implementation. The ISDN private branch exchanges [3.16], which are connected to the public ISDN and, if required, interconnected via basic- or primary-rate accesses (see Sect. 4.1.5), bring the public ISDN to the users of an ISDN PBX and offer them all the basic office communication services such as text, data and facsimile transmission, in addition to the telephone service. Local area networks (LANs) are used for high-speed data packet exchange between workstation systems, computers and centralized high-performance terminals located within a limited "local" area [3.17]. Unlike the ISDN PBXs, which employ circuit switching, the LANs are based on the packet switching principle and use access protocols controlled on a decentralized basis. As a large number of users access the same transmission medium, LANs require high-bandwidth transmission systems. "Gateways" allow interoperation between the LANs and ISDN PBXs by performing the required signaling adaptation functions and transmission rate conversions. The LAN as an entity is therefore seen by the ISDN as an ISDN terminal with the S- or T-interface (see Sect. 4.1.5). This means that the functions of a matching unit (see Sect. 4.1.4) are implemented in the gateway (Fig. 3.8).

The interconnection of LANs over comparatively large distances and the setting-up of Metropolitan Area Networks (MANs) incorporating network components of the public network constitute further forms of private network organization. Configurations of this kind require dedicated and switched connections with a bandwidth of 2 Mbit/s and above, which can be implemented as part of the evolution of 64-kbit/s ISDN to broadband ISDN.

3.7 Numbering

ISDN numbering is based on the telephone network numbering plan [3.18]. An important difference is that, in the ISDN numbering plan, the maximum number of digits in the numbers is increased from 12 to 15 (see Sect. 4.3.3.2) [3.19]. Of these, only the first three are reserved for the unchanged 1- to 3-digit country codes, the remaining digits form the national number and are available for the network identifier, the local area code, the subscriber number and – in Germany, for example, – for more specific terminal selection in the case of bus configurations. The digits required for traffic discrimination, such as 0 for national or 00 for international long-distance traffic, do not form part of the ISDN number. The network identifiers allow gateways to service-specific networks or different ISDNs within a country to be addressed [3.20].

The ISDN numbering plan allows for the fact that a comparatively large number of digits of the destination country's national directory number must be evaluated in the originating country. For this purpose, some modifications must be made to the call-processing facilities of the existing worldwide telephone network. With this in mind, the transition date for changing over to the full potential of the ISDN numbering plan has been fixed as 31.12.1996 [3.21]. In addition to the ISDN number, the ISDN provides the ISDN subaddress. This comprises up to 40 digits and is transmitted transparently from the calling to the called subscriber during call setup. The ISDN subaddress and number are separated by an identification code. The subaddress allows the sub-components of the called subscriber to be addressed more precisely than under the ISDN number (see Sect. 4.3.3.2). As the ISDN offers several communication services, a criterion is required during call setup to indicate which service the subscriber wishes to use. For this purpose, a service identification code is employed which is automatically transmitted by the terminals if required. This enables the network to ensure that the connection is set up via suitable paths, e.g. via an end-to-end 64-kbit/s link, and at the called subscriber end it is possible to check whether suitable terminals are available for the desired service (compatibility check) (see Sect. 4.3.3.3).

This check ensures full compatibility in the case of the teleservices (see Sect. 2), but in the case of the bearer services merely indicates the presence of a terminal of the appropriate speed category, i.e. not necessarily the compatibility of the link protocols also.

3.8 Implementation Strategies

Telephone networks worldwide are currently being converted from analog to digital transmission and switching – *Integrated Digital Network* – because this confers economic advantages. Since it is desirable for a number of reasons (transmission quality, cost-effectiveness) to avoid several analog/digital conversions within a connection, network digitization is mostly aimed at providing wholly 64 kbit/s connections between originating and destination exchanges at an early date. As a further conversion measure within the trunk network, signaling associated with the traffic channel will be replaced by the more powerful Signaling System No. 7 which employs central signaling channels (see Sect. 6.3). Once the subscriber lines are also converted to digital transmission, all the conditions will be created for extending the digital telephone network into the ISDN.

3.8.1 Overlay Network and Cell Approach

Two basic implementation strategies can be identified for the transition from the existing network to the ISDN: the overlay network and the cell approach [3.22].

The aim of implementation by the *overlay network strategy* is to make the new technology available in a large area even in the initial phase, and with as little new equipment as possible. At suitable points in the network new exchanges will be built which are connected entirely via digital transmission systems and which exchange their switching information via Signaling System No. 7. Interworking with the existing network will be implemented with as few gateways as possible in order to minimize expenditure on conversion equipment, e.g. for analog/digital and signaling conversion. Calls from ISDN subscribers to subscribers in the existing network will be routed as far as possible within the ISDN, while calls in the opposite direction will be transferred to the ISDN at the nearest gateway (Fig. 3.9). As well as replacing complete exchanges, the introduction of new switching systems can also take the form of system expansions. In this case they will primarily be used to cover the demand for ISDN subscriber lines.

The disadvantages of the overlay network are the initially low usage of the overlay links, the limited traffic routing possibilities and problems of operation and maintenance, as a network with a large extent but comparatively low density must be operated and maintained by the operating centers of the existing network.

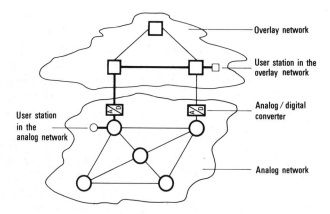

Fig. 3.9. Network Implementation by the Overlay Method.
────── Connection between user stations in the overlay network and in the analog network

The main feature of the *cell approach* is that it concentrates the introduction of the new network components in a defined area, perhaps even providing complete network conversion within that area before new switching and transmission systems are introduced in other areas. This approach is more theoretical than practical in nature, as even within a limited area it is generally unrealistic to replace the entire existing network at a stroke. Furthermore, with this implementation strategy the subscribers of one region would be given preferential treatment over the others.

3.8.2 Pragmatic Implementation Strategy

The overlay network and the cell approach represent two basic implementation strategies which, however, do not take sufficiently into account, or even ignore, a number of practical factors some of which may be specific to particular countries. These include:

- the scale of conversion from analog to digital transmission and switching and the geographical distribution of the existing digital network equipment,
- the extent and location of the demand for new services,
- the condition and age structure of the existing network equipment,
- the space available for installing new network equipment in the buildings,
- the existing infrastructure for operation and maintenance of the network and the qualification level of the personnel involved,
- the financial resources of the network carrier.

As the two implementation concepts are not incompatible, network implementation will mostly be a combination of the overlay network and cell approach, taking into account the factors listed above. To give an example of this approach to network conversion, known as the *pragmatic implementation strategy*, the steps by which the Federal Republic of Germany intends to bring about the transition from the analog to the digital telephone network and to the ISDN are described below [3.23].

- Since the mid-seventies, digital transmission systems have been used mainly in the short-haul network, i.e. the trunks connected to primary centers. By the mid-eighties, an initial phase involving 2 Mbit/s and 34 Mbit/s transmission systems (30 or 480 64-kbit/s channels) was operational at this level. In the long-haul network, transmissions systems operating at 140 Mbit/s and 565 Mbit/s (1920 or 7680 64-kbit/s channels) are also being introduced.
- Since 1985, digital local and long-distance exchanges have been installed, and since the end of the eighties only digital switching systems have been acquired. For the *long-distance exchanges*, network introduction from "top to bottom" is planned, i.e. digital switching equipment will first be installed in the largest transit exchanges in the major cities. These exchanges will be given "digital expansion units". Introduction of digital *local switching* is generally intended to be coordinated with long-distance switching, i.e. each time there is a digital

expansion in the associated long-distance switching system. It is expected that by 1990 approximately 20% of the long-distance exchanges will be equipped with digital systems, and that in 1995 approximately 20% of the lines in the Federal Republic, i.e. approximately 6 million lines, will be connected to digital local exchanges.

- At the start of 1989, ISDN was officially inaugurated in the Federal Republic of Germany. All the newly installed exchanges are equipped for ISDN, and from 1990 onwards the digital exchanges cut over at an earlier date are being upgraded accordingly. The local networks with ISDN capability are interconnected via digital transmission paths, thereby allowing ISDN traffic throughout Germany.

From the above description the following conclusions can be drawn for the ISDN network configuration in the Federal Republic of Germany:

- Even in the introductory phase the ISDN will not form a separate overlay network, i.e. a separate network superimposed on the telephone network which has access to the analog parts of the telephone network via *central* gateways. It will rather be a network embedded in the digital telephone network, in which only digital network equipment will be used for connection set-up between two ISDN subsribers (Fig. 3.10). Unlike an implementation strategy based on the overlay principle, this less rigid approach is expected to provide greater economic advantages, especially during the transitional phase [3.14, 3.24].
- As the ISDN is developing out of the telephone network, its network structure will largely be based on that of the present-day telephone network (Fig. 3.2). The stored program controlled exchanges enable more flexible

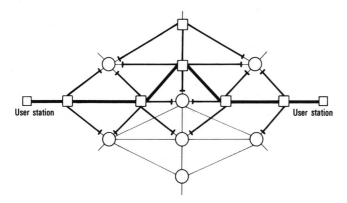

Fig. 3.10. ISDN Connection in a Hybrid Analog/Digital Network.

○	□	Exchange
———	———	Trunk circuit
———	ISDN connection
	Analog/digital converter

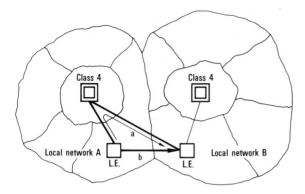

Fig. 3.11. Extended Traffic Routing Options With Stored-Program Controlled Exchanges.
a   Shortest possible connection in the analog telephone network,
b   Possible connection with stored-program controlled exchanges,
L.E. Local Exchange

traffic routing; for example, high usage trunk groups can be installed between local exchanges in adjacent local networks (Fig. 3.11), which was not possible using electromechanical exchange equipment. The greater flexibility of traffic routing also enables the number of hierarchical levels in the long-distance network to be reduced.

• The ISDN will radiate out from the large local networks which have within their area transit exchanges of the upper hierarchical levels. These are largely interconnected via digital transmission paths; in this way a wide-area network reaching the major centers of population can be provided at an early stage. As it is there that the majority of medium and large private branch exchanges are located, the potential users of the ISDN will be able to have an ISDN line without delay.

3.8.3 Satellite Links in the ISDN

One of the basic conditions of the ISDN is an all-digital connection from subscriber to subscriber. For ISDN exchanges located at great distances from each other there may be no all-digital transmission paths available in the terrestrial network in the introductory phase. In these cases, an alternative is connection via satellite. Besides the INTELSAT satellites for intercontinental traffic which have been in operation since the sixties, national satellite projects are operating or being planned in several countries. These provide channels with bit rates 64 to 2048 kbit/s for transmitting speech, text, data and pictures and are being considered for connecting ISDN exchanges or ISDN private branch exchanges separated by great distances.

The disadvantage of satellite links is the longer signal delay compared with terrestrial connections (approximately 260 ms for the link from earth station to satellite to earth station) [3.25]. For text and data connections, in which control instructions and acknowledgements must be transmitted in both directions in the transfer phase, this may considerably prolong the holding time. In telephone connections, the long signal delay makes conversation more difficult (cf. Sect. 7.7.3); as in the analog telephone network, echo suppressors or echo cancelers will be necessary.

Due to residual orbit ellipticity and to the gravitational pull of the moon, signal delay variations occur in satellite connections which cannot be fully compensated by correcting the satellite orbit. These signal delay variations, of the order of a millisecond, must be equalized in buffer memories.

Thus, although satellite links form a flexible supplement to the terrestrial trunk network, their use in digital communication networks requires a number of additional measures to ensure a transmission quality comparable with that of terrestrial links.

3.9 ISDN Implementation Schedules

Virtually all the telecommunication administrations and operating companies in the western industrialized world have defined objectives as to the methods and schedules by which the transition from the analog to the digital telephone network and then to the ISDN is to be completed [3.26].

As far as the channel structure on the subscriber line is concerned, most operating companies have opted for a system using two 64 kbit/s basic channels and one 16 kbit/s signaling channel ($2 \times B + D$). Different solutions with net bit rates of 80 and 88 kbit/s were initially adopted in Great Britain and Japan [3.27, 3.28], as these countries had already decided to introduce ISDN at an early date. However, both have since then changed over to the internationally standardized arrangement of two 64-kbit/s basic channels and one 16-kbit/s signaling channel.

The user potential for the ISDN will initially come mainly by business users, most of which have ISDN private branch exchanges or use Centrex services. For this reason, in many countries, when the ISDN is implemented the primary rate access will be provided as well as the ISDN basic access. Different strategies will be pursued by the carriers or operating companies in terms of interworking with existing dedicated networks and of provision of packet-switched services in the ISDN. Common to these strategies is the desire to provide a range of services at least equivalent to those offered by the dedicated networks.

In Europe, at the end of 1986 the Council of the European Community adopted a detailed recommendation on the coordinated introduction of a single version of the ISDN. The recommendation has set deadlines for the specification and implementation of a limited number of ISDN services.

As a result of that recommendation, the "Memorandum of Understanding (MOU) on the Implementation of European ISDN service by 1992" has been signed in 1989 by 22 telecommunication operators of 18 European countries; others joined later [3.29]. The MOU commits its signatories to provide a minimum set of services based on uniform standards by the end of 1993. This set of services includes a circuit-switched 64 kbit/s bearer service that is transparent to user traffic, and the 3.1 kHz audio bearer service. In addition to the basic channel support, a number of supplementary services are also to be provided. These are:

– calling line identification presentation,
– calling line identification restriction,
– direct dialing in,
– multiple subscriber number,
– terminal portability.

Furthermore the MOU commits each MOU signatory to provide ISDN interconnections to all the networks of the other signatories.

4 Subscriber Access

Compared with the digital telephone network in which analog signals are transmitted to the user equipment, the essential technical innovation of ISDN is the digitization of the subscriber line (see Fig. 3.4). The majority of the internationally agreed ISDN specifications (see list of CCITT Recommendations of the I Series in the Annex to this book) on which the following remarks are based therefore deal with subscriber access.

The main aspects of user access are the configuration of the user station (Sect. 4.1), the user-network interfaces within the user station (Sect. 4.2) and user signaling (Sect. 4.3).

4.1 Configuration of the User Station

4.1.1 Functional Groups of the User Station

Figure 4.1a shows the configuration of a user station as specified in CCITT Recs. I.410 [4.1] and I.411 [4.2]. One or more terminal equipments (TE) are connected to the network termination (NT). The terminal equipments can be of the same type, e.g. a number of telephones, or there may be a combination of different types such as voice and non-voice terminals.

The network termination NT physically connects the terminal equipment to the subscriber line and enables the latter to be used jointly by several terminals. For these two functions the NT is divided into two functional groups, NT1 and NT2, as shown in Fig. 4.1b: NT1 provides the physical connection to the subscriber line, NT2 enables it to be used by several terminals.

The functional group TE can either be a terminal equipment type 1 (TE1) specially designed for the ISDN (Fig. 4.1b) and connected directly to the interface at reference point S, or a terminal equipment type 2 (TE2) with a conventional interface connected via a terminal adaptor (TA).

Reference points between the functional groups have been defined: reference point T between network terminations NT1 and NT2, reference point S between network termination NT2 and the terminal equipments TE, or more precisely TE1 or TA. A standardized physical interface can, but does not always have to be provided at these reference points (cf. Sect. 4.1.3).

Depending on national or network-specific regulations, the responsibility of the network operator ends at reference point S, T or U (cf. Fig. 4.1). If it ends at

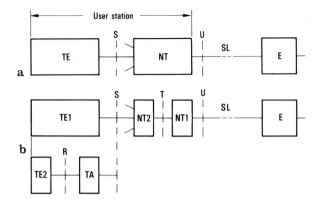

Fig. 4.1a, b. Configuration of the User Station.
SL subscriber line, NT, NT1, NT2 network termination, R, S, T, U reference points, TA terminal adaptor, TE terminal equipment, TE1 terminal equipment with ISDN interface, TE2 terminal equipment with conventional interface, E exchange

reference point S, the network operator is responsible for network terminations NT2 and NT1. If it ends at T, he is responsible for network termination NT1 only and if it ends at U, he is responsible for neither NT2 nor NT1. The particular reference point (S, T or U) at which the responsibility of the network operator ends, is both the point at which the network operator supplies a defined service – access to the communication services (cf. Sect. 2) – and the point up to which he accepts responsibility for maintenance. In those cases in which the network operator is responsible for network terminations NT2 or NT1, national or network-specific rules define whether this point is located directly at the NT output or whether the installation as far as the sockets for user terminals (cf. Fig. 4.8) also falls within the network operator's sphere of responsibility.

To provide the user with universal access to the ISDN communication services, the interface at reference points S and T is internationally standardized. For reference point U, on the other hand, only different national specifications have so far been agreed; further national variants and perhaps international standards may emerge (cf. Sect. 4.1.2). The standard for reference points S and T covers not only the mechanical and electrical specifications (Sect. 4.2), but also the specifications for the operating procedure (Sects. 4.2 and 4.3) with the aim of allowing problem-free interconnection of terminal equipments and networks of different origin.

For this purpose the same set of specifications is used for both reference points S and T. If no special NT2 functions (e.g. internal traffic) are required, functional group NT2 can be reduced to a "zero NT2" (cf. Sect. 4.1.3); a terminal equipment designed for reference point S can therefore also operate at reference point T.

Between TE2 and TA lies reference point R at which a conventional interface is normally implemented (cf. Sect. 4.1.4), e.g. in accordance with the CCITT

V.-Series Recommendations (see Annex), CCITT Recs. X.21 [4.3] or X.25 [4.4] or the RS232 interface defined by the EIA (Electronic Industries Association, Washington D.C., USA). In order that an ISDN terminal equipment of type TE1 and a combination of TE2 and TA (see Fig. 4.1b) can be handled uniformly by the network (e.g. in terms of addressing), there is provided a separate terminal adaptor TA for each TE2, at least from the functional viewpoint.

4.1.2 Network Termination NT1

The network termination NT1 converts the signals at reference point T to signals suitable for transmission on the subscriber line and vice versa (see Sect. 7.4.3).

As subscriber lines can differ greatly from country to country and even within a country (e.g. in terms of length, cable characteristics, branches etc.), the transmission method for the subscriber line has not yet been internationally standardized (cf. Sect. 4.1.1); network operators define their own specifications for their areas. Due to the nature of functional group NT1, the other functional groups of the user station (i.e. NT2, TE1, TA, TE2) operate independently of the transmission method on the subscriber line.

The network termination NT1 has another important role in fault location. If malfunctions occur, it is necessary to establish, from the switching equipment, the location of the fault in order to enable maintenance personnel to be efficiently deployed. The following test loops (Fig. 4.2, cf. Sect. 3.3), some of which may be standardized by the CCITT and which can be controlled if required from the switching equipment [4.5], may be used for this purpose:

- With test loop c, a switching equipment self-test is performed before each connection set-up
- Test loop b indicates whether the subscriber line is properly terminated (by an NT1) at the user end so that the switching equipment can detect any fault present on the subscriber line.
- With test loop a, the switching equipment can ascertain whether the signals are being correctly transmitted in both directions on the subscriber line; under control of the switching equipment, the NT1 loops some or all of the information sent by the switching equipment back to the switching equipment which then determines whether the bit error rate is within the permitted tolerance range. Depending on how much information is looped back (e.g. all channels or only one channel, see Sect 4.2.1.1) with the test loop activated, the

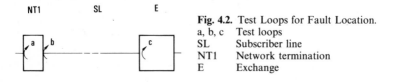

Fig. 4.2. Test Loops for Fault Location.
a, b, c   Test loops
SL   Subscriber line
NT1   Network termination
E   Exchange

line is either not available at all or only available to a limited extent to the user.

In addition to this test method which, though effective, impairs operation, the network or the user can if necessary use in-service test methods; for example, suitable data protection information can be added to the transmitted information, enabling the user to ascertain the bit error rate.

Also under discussion are tests which can be initiated by the terminal equipments, e.g. a facility for requesting the network to test the line, or test loops in which information sent by a terminal equipment is looped back to it by the local NT or by the distant NT.

4.1.3 Network Termination NT2

The main function of network termination NT2 is to allow joint use of a network access by more than one terminal equipment (cf. Fig. 4.1).

The number of terminal equipments, their spatial arrangement and also, therefore, the implementation of the NT2 may vary considerably (Fig. 4.3).

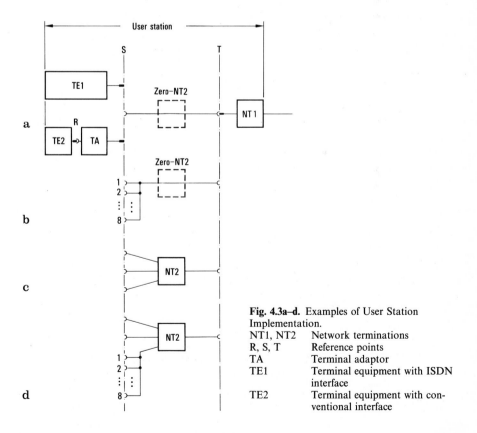

Fig. 4.3a–d. Examples of User Station Implementation.

NT1, NT2	Network terminations
R, S, T	Reference points
TA	Terminal adaptor
TE1	Terminal equipment with ISDN interface
TE2	Terminal equipment with conventional interface

Figure 4.3a shows the user station referred to in Sect. 4.1.1 with one terminal equipment and a "zero NT2". The "passive bus" (see Sect. 4.2.2.1) shown in Fig. 4.3b is another example of a "zero NT2"; up to eight terminal equipments can be connected in this way. This configuration is only possible with the basic access (see Sect. 4.2.1.2).

However, the NT2 can also be very powerful equipment, such as a private branch exchange which concentrates the traffic of many terminal equipments (Fig. 4.3c) and provides the terminal equipments with additional facilities (e.g. internal traffic). An NT2 of this type can itself provide the above mentioned "passive bus" (Fig. 4.3d).

The connection between NT2 and NT1 may be implemented in various ways (Fig. 4.4a to d).

The normal configuration used for many applications (especially where the user traffic requirement is low) consists of an NT2 connected to a single NT1 (Fig. 4.4a). To meet more exacting requirements in terms of traffic and availability, an NT2 can have multiple connections to NT1 units (Fig. 4.4b and c).

For economic reasons it is also possible to combine various functional groups to form a single physical entity (Fig. 4.4d and e).

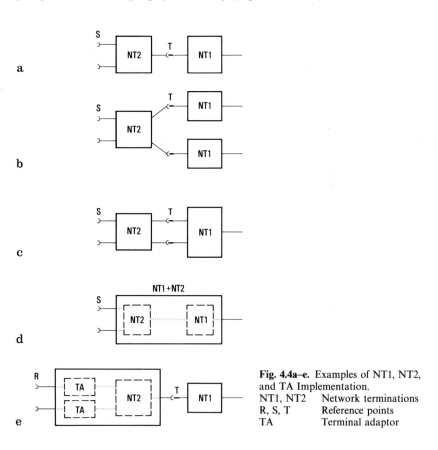

Fig. 4.4a–e. Examples of NT1, NT2, and TA Implementation.

NT1, NT2	Network terminations
R, S, T	Reference points
TA	Terminal adaptor

4.1.4 Terminal Adaptor TA

From the economic and organizational standpoint, the possibility of connecting conventional terminals to the ISDN via a terminal adaptor TA facilitates access to the ISDN for the user. The latter should be able to reach his communication partners with both conventional and new terminal equipments, even if his partners are still connected to dedicated service networks (cf. Chap. 2).

The terminal adaptors are normally connected to the interface at reference point S (Fig. 4.3); one or more terminal adaptors can also be combined with the NT2 to form a single entity (Fig. 4.4e) which is then connected to the interface at reference point T.

In drafting the specifications for reference point S, care was taken to ensure that the main CCITT interfaces, e.g. the interfaces defined in CCITT Recommendations X.21 [4.3], X.25 [4.4] and in the V.-Series (see Annex), could be adapted without difficulty. The specifications for the terminal adaptors are given in CCITT Rec. I.461 [4.6] for X.21, in I.462 [4.7] for X.25 and in I.463 [4.8] for the V.-Series interfaces. The interface corresponding to the access point for an analog subscriber line in the conventional telephone network ("t/r i.e. tip and ring interface") was also taken into account. The extent to which other interfaces, too, can be adapted must be checked for each individual case.

4.1.5 Connecting Private Networks

The possibilities for connecting private networks to the ISDN can be illustrated by examining the connection options for private branch exchanges (Fig. 4.5).

There are three alternative methods of connecting private branch exchanges (Fig. 4.5): connection at reference point S, connection at reference point T and

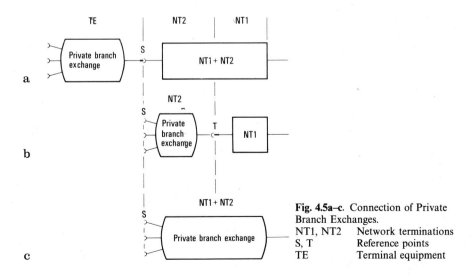

Fig. 4.5a–c. Connection of Private Branch Exchanges.

NT1, NT2	Network terminations
S, T	Reference points
TE	Terminal equipment

direct connection to the subscriber line at reference point U. With connection at reference point S (Fig. 4.5a) the private branch exchange is perceived by the network to be connected like a terminal equipment; in this case NT1 + NT2 does not need to provide any NT2 functions (cf. zero NT2 in Sect. 4.1.3). With connection at reference point T (Fig. 4.5b) the private branch exchange acts as an NT2; with direct connection to the subscriber line (Fig. 4.5c) it acts like an NT1 and NT2.

As the same interface specifications apply to reference points S and T (cf. Sect. 4.1.1), the first two alternatives are technically identical. They may, however, differ as regards the user facilities provided by the network. An example of this is direct dialing-in using the ISDN number (see Sect. 4.3.3.2): as CCITT Rec. I.330 [4.9] currently stands, the network address range corresponding to the ISDN number only extends as far as reference point S and is not designed for sub-addressing within terminal equipments (for this purpose there is the *ISDN subaddress*; see Sect. 4.3.3.2). A private branch exchange featuring direct dialing-in (using the ISDN number) must therefore be connected as an NT2 unit at reference point T.

Direct connection to the subscriber line can give cost benefits as compared with the first two alternatives, especially if a standard interface between NT1 and NT2 is not used; however, the private branch exchange is then subject to the interface conditions (e.g. transmission methods) of the subscriber line.

In the user domain it is possible to install a complex private network, e.g. a network which can consist of a combination of interlinked private branch exchanges, of local area networks (LANs) meeting ISO Draft International Standard 8802 [4.10] and communication components of a private computer network (Fig. 4.6). A network of this type can provide ISDN and other communication services (cf. Sect. 3.6.2), and terminal equipments with ISDN interfaces can be attached as well as terminal equipments with other interfaces. Such complex networks can be connected to the public ISDN in the same way as private branch exchanges (cf. Fig. 4.5).

A private network not implemented in accordance with the ISDN concept can in some circumstances be endowed with ISDN capability by means of

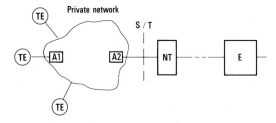

Fig. 4.6. Connection of Private Networks.

A1	Adaptor for connection of ISDN terminal equipments	TE	Terminal equipment
A2	Adaptor for connecting the private network to the ISDN	NT	Network termination
		S, T	Reference points
		E	Exchange

suitable terminal adapters within the private network: terminal adaptor A1 (Fig. 4.6) can enable ISDN terminal equipments to be connected, terminal adaptor A2 (Fig. 4.6) the connection of the private network to the (public) ISDN.

The ISDN addressing mechanisms (see Sect. 4.3.3.2) are sufficient to permit individual addressing of the terminal equipments connected to the private network, and also to interlink several local private networks with the aid of the (public) ISDN to form one large private network.

4.2 User-Network Interfaces

4.2.1 Preliminary Remarks

The specifications relating to the user-network interface have a special significance in communication networks: nationally and internationally, they define the demarcations between terminal equipments and network components.

The user-network interface specified for the ISDN is based on the concept of the ISDN as a universal communication network intended to enable any person worldwide to communicate by any method – using voice, text, data and picture simultaneously – , a network to which it must be possible to connect both specialized and multifunction terminal equipments.

All communication over the ISDN is based on digital links, each with a certain transmission capacity. A user's transmission capacity requirement depends on the type and number of communication services used at the same time, and may vary considerably from user to user; for example, one user may wish to connect an equipment serving as a combined high-quality video telephone set and videotex terminal, another may require a conference facility, a third may need several terminal equipments such as telephones and data terminals and a fourth a large private branch exchange with a high traffic handling capacity.

In order to cover such a variety of requirements without an excessively large number of interface variants, the number of access types defined has been kept to a minimum (see Sect. 4.2.1.2) with transmission capacity levels spread as widely as possible. For the two access types currently specified, the basic access and the primary rate access, the net transmission capacity differs by at least a factor of 10.

In order to adapt the transmission capacity provided by the network to suit actual user requirements, access arrangements of the same or different types can be "connected in parallel"; in addition, the network may not make the full transmission capacity available at the interface (see Sect. 4.2.1.3).

The characteristics of the user-network interface are defined in detail for the basic access and primary rate access. These include specifications for the functional, electrical and mechanical characteristics (see Sects. 4.2.2 and 4.2.3) and for the operational procedures (see Sect. 4.3). The basic specifications described below for the user-network interfaces – channel types, access codes and interface structures – are based on CCITT Rec. I.412 [4.11].

4.2.1.1 Channel Types

The net transmission capacity available at the user-network interface is sub-divided in a specific manner – depending on the type of access (see Sect. 4.2.1.2) – into one or more *traffic channels* (e.g. B channels) and normally an additional *signaling channel* (D channel). In special cases (see Table 4.3) the interface structure does not contain a signaling channel, or the active signaling channel can be routed via a different access.

The traffic channels are used by the network to provide circuit-switched and normally also packet-switched communication modes; the D channel is used for communication between user terminal and network, in other words for signaling (cf. Sect. 4.3).

Insofar as the signaling channel has spare capacity in addition to that required for signaling, it can be used for transmitting packet-mode data (also for teleaction purposes) in accordance with the relevant national or network specifications. Signaling and packet-mode data are interleaved for transmission on the same channel, signaling having priority.

Types of Traffic Channel

The traffic channels currently defined are the three types with different transmission capacities listed in Table 4.1.

The most important channel type is the 64 kbit/s channel: the B channel, also known as the basic channel. Its bit rate is based on the 8-bit PCM encoding (octet structure) of the telephony signal (see Sect. 7.2.1). The 384 kbit/s H0 channel and two variants of the H1 channel operating at 1920 kbit/s (H12) and 1536 kbit/s (H11) are defined as higher-capacity types of the traffic channel.

In order to be able to provide the basic octet structure (see Sect. 7.2.1) for the PCM-encoded telephony signal in all these channel types without using complicated ancillary equipment, the sender may structure the information in 8 kHz units for all channel types. An information structure of this kind is retained all the way through the network to the receiver. One 8 kHz unit consists of as many bits as can be transmitted within 125 μs (corresponding to 8 kHz): i.e. 8 bits (octet) for 64 kbit/s channels, 48 bits for 384 kbit/s channels, etc.

Also under consideration are channels with a lower capacity than the B channel, so-called "sub-rate channels" of 8, 16 and 32 kbit/s. Using such channels, the network could make more connections to different destinations

Table 4.1. Types of Traffic Channel

Channel designation	Bit rate kbit/s
B	64
H0	384
H11	1536
H12	1920

Table 4.2. Types of Signaling Channels

Channel designation	Bit rate kbit/s
D_{16}	16
D_{64}	64

simultaneously on one access than on a B-channel basis; however, these connections have less information-carrying capacity than 64 kbit/s connections, and the switching equipment becomes more costly. Another problem with sub-rate channels is that they may lead to a proliferation of classes.

Types of Signaling Channel
The signaling channel is called the D channel (Table 4.2). Depending on the type of access (see Sect. 4.2.1.2), the D channel has a bit rate of 16 or 64 kbit/s. The standardized D channel protocol LAPD (see Sect. 4.3.4) is used as the data link layer protocol. The layer 3 protocol is described in Sect. 4.3.5.

4.2.1.2 Access Types and Interface Structures

Access Types
Two access types are currently defined for reference points S and T: the *basic access* and the *primary rate access*.

With the basic access, a signal with a total bit rate of 192 kbit/s is used in both directions; the net bit rate then available for the traffic channels and the signaling channel is 144 kbit/s (cf. Sect. 4.2.2.4 and Fig. 4.10).

With the primary rate access, a signal with a total bit rate of 2048 or 1544 kbit/s is used in both directions; the net bit rate is 1984 kbit/s or 1536 kbit/s (see Sect. 4.2.3 and Fig. 4.14).

Subdivision of the net bit rate into channels produces defined *interface structures* at the interface.

Interface Structures
Table 4.3 shows the interface structures specified for the two above-mentioned access types. They are identical for both transmission directions; consequently all channels can be used simultaneously in both directions.

In the case of the basic access, there is only one interface structure with two B channels (64 kbit/s each) and one D channel (16 kbit/s) for signaling (Fig. 4.7).

Several interface structures are defined for the primary rate access, the B-channel – , H0-channel and H1-channel interface structures and the "mixed" interface structure in which the net bit rate can be subdivided into any combination of B and H0 channels (cf. Sect. 4.2.3). The bit rate of the signaling channel is always 64 kbit/s.

If appropriate specification exists nationally or in respect of individual networks, the signaling for a primary rate access may be routed via the 64 kbit/s

Table 4.3. Interface Structures at Reference Points S and T

	Basic access	Primary rate access	
Net bit rate	144 kbit/s	1984 kbit/s	1536 kbit/s
B-Channel structures	$B + B + D_{16}$	$30 \cdot B + D_{64}$	$23 \cdot B + D_{64}$
	—	$30 \cdot B^a$	$24 \cdot B^a$
H0-channel structures	—	$5 \cdot H0 + D_{64}$	$3 \cdot H0 + D_{64}$
	—	$5 \cdot H0^a$	$4 \cdot H0^a$
H1-channel structures	—	$H12 + D_{64}$	—
	—	$H12^a$	$H11^a$
Mixed structures	→	$n \cdot B + m \cdot H0 + D_{64}$	$n \cdot B + m \cdot H0 + D_{64}$
	—	$n \cdot B + m \cdot H0^a$	$n \cdot B + m \cdot H0^a$

D_{16} 16 kbit/s D-channel
D_{64} 64 kbit/s D-channel
[a] The associated signaling channel, if required, is routed via a different access.

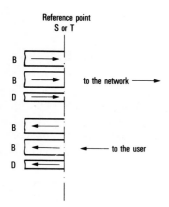

Fig. 4.7. Interface Structure of the Basic Access
B general channel with 64 kbit/s
D auxiliary channel (16 kbit/s) e.g. for user signaling

signaling channel of another primary rate access. (With the H1-channel interface structure the 16 kbit/s signaling channel of a basic access can also be used in exceptional cases.) Assignments are made when the access arrangements are set up and can be changed if required. The signaling channel capacity which is released is available for traffic channels in the 1544 kbit/s variant, normally remaining reserved in the 2048 kbit/s variant.

4.2.1.3 Operation of Traffic Channels

With certain access arrangements or generally with a specific access type, the network may operate some traffic channels either not at all or only with restricted transmission capacity. These channels are nevertheless incorporated with their full bit rate in the interface structure (cf. Sect. 4.2.1.2) at the user-network interface.

If it is economically desirable, the network operator may thus offer the user the available transmission capacity in a more graduated manner (cf. Sect. 4.2.1), and may tailor the equipment associated with the access to suit the user's actual requirements. For example, in the case of the primary rate access with B-channel interface structure, the network may only operate some of the many B channels contained in the interface structure.

During the introduction phase, in some countries one of the two B channels of the basis access will only transmit over the subscriber line at 8 or 16 kbit/s and not at the full bit rate (64 kbit/s) (see Sect. 3.9). Naturally, the network can only assign correspondingly restricted network connections to such a channel. In the case of interworking with a 64 kbit/s channel suitable rate adaption methods have to be used (standardized rate adaption methods are described in CCITT Recs. I.460 [4.12], I.461 [4.6], I.463 [4.8] and I.464 [4.13] (cf. Sect. 4.4)). After an introduction phase all network operators (as far as is currently known) intend to operate the basic access at full capacity.

4.2.2 User-Network Interface for the Basic Access

The user-network interface is the means by which the user terminals gain access to the channels. The timing for information exchange is supplied by the network (see Sect. 7.6.2). The electrical characteristics (see Sects. 4.2.2.2 and 4.2.2.6) and the procedural characteristics (see Sects. 4.2.2.3 to 4.2.2.5) are designed for specific, defined model configurations (see Sect. 4.2.2.1). This does not preclude other configurations yet to be standardized. The standardized specifications for layer 1 (cf. Sect. 4.3.1) of the basic user-network interface are detailed in CCITT Rec. I.430 [4.14].

4.2.2.1 Reference Configurations

The definition of the electrical and procedural characteristics is based on three reference configurations (Fig. 4.8): the point-to-point configuration, the passive bus and the extended passive bus. None of these configurations require NT2 functions (zero NT2, cf. Fig. 4.3a and b). However, the two passive bus configurations require the network to be capable of controlling several (e.g. up to eight) terminal equipments on one access.

Each terminal equipment TE (i.e. TE1 or TA + TE2, cf. Fig. 4.1b and Sect. 4.1.1) is attached via a connecting cord CC_{TE} up to 10 m long and an internationally standardized eight-pin plug (ISO 8877 [4.15]) to a socket at the interchange cable IC (Fig. 4.8). Any stub connecting the socket to the interchange cable must not exceed one meter in length. The network termination NT is connected to the cable either permanently or via the same type of plug as the terminal equipments. As the electrical characteristics of the connecting cord may have a considerable effect on transmission performance, especially in the passive

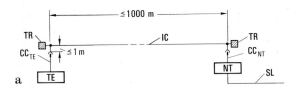

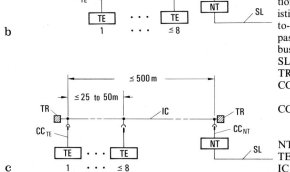

a

b

c

Fig. 4.8a–c. Reference configuration for Specifying the Characteristics for the Basic Access. **a** Point-to-point configuration; **b** (short) passive bus; **c** extended passive bus.

SL   Subscriber line
TR   Terminating resistor
CC_{NT} Connecting cord for network termination ($\leq$ 3 m)
CC_{TE} Connecting cord for terminal equipments ($\leq$ 10 m)
NT   Network termination
TE   Terminal equipment
IC   Interchange cable

bus arrangement, these characteristics have recently been standardized by CCITT. An extension cord up to 25 m long is allowed for cord CC_{TE} in the point-to-point configuration, provided this does not increase the overall attenuation to over 6 dB.

Normally two wire pairs are used at the interface; these can be used for transmission (one wire pair for each direction) and optionally for power-feeding of terminal equipments from the NT (see Sect. 4.2.2.6 and Fig. 4.13). The advantages of two wire pairs include cost, operational reliability, the range of the passive bus and ease of implementation.

In special cases (depending on the network operator's specifications) the NT can power the terminal equipments via an additional third wire pair (see Sect. 4.2.2.6 and Fig. 4.13). The eight-pin plug arrangement allows yet another wire pair to be used; the latter is not required by ISDN standards, but could be used e.g. for mutual feeding of two directly interconnected terminal equipments.

The interchange cable IC is subject to no special requirements; it will normally consist of two unshielded, balanced, twisted pairs. For the transition from an analog line to the ISDN, normally the subscriber line (cf. Sect. 7.4.3) and frequently also the wiring on the user premises can continue to be used. The two wire pairs can be combined with other wire pairs in one cable. The complete cable run including sockets is purely "passive" in all three configurations, i.e. it contains no amplifying, storing or processing functions. To prevent reflections, a terminating resistor is required at each end of the cable (Fig. 4.8); the terminating resistor at the NT end can be incorporated in the NT.

In all the configurations, the range is limited by the signal delay and the attenuation. Both these variables depend on the connecting cable used: Fig. 4.8 shows possible ranges using typical cabling by way of illustration.

In the point-to-point configuration (Fig. 4.8) the attenuation – measured at 96 kHz – must be less than 6 dB. Specific rules apply to the other two configurations (see CCITT Rec I.430 [4.14].

In order to enable the NT to loop back the D-channel information transmitted by the terminal equipments with correct timing in the D-echo channel (cf. Fig. 4.10 and Sect. 4.2.2.3), the round trip delay NT-TE-NT must be less than 42 µs (approximately eight times the bit period). This assumes that the NT needs a maximum of 10.4 µs (twice a bit period) for this loopback process. This condition is normally uncritical for all configurations.

In the passive bus arrangement (Fig. 4.8b), terminal equipments can be connected at any point – e.g. one nearby the NT, another at the very end of the passive bus – and can transmit simultaneously. The signal delays of the individual terminal equipments vary with distance. In order to ensure that the signals transmitted by several terminal equipments in the same frame can still be interpreted by the NT, it is necessary to limit among other things (cf. Sects. 4.2.2.2 to 4.4.4.4) the permitted round trip delay and hence also the range of the passive bus. The round trip delay NT-TE-NT – disregarding the frame offset between TE input and TE output (10.4 µs) – must therefore be less than 3.6 µs (approximately 70% of a bit period) for all terminal equipments. If in the passive bus arrangement the phase readjustment required for the other two configurations is not deactivated in the NT, this figure is reduced to 2.6 µs (approximately 50% of a bit period).

With the extended passive bus (Fig. 4.8c), the reception conditions for the NT are less favorable than with the passive bus (Fig. 4.8b). Consequently, the NT-TE-NT round trip delays for all the terminal equipments must vary by no more than 2.0 µs (approximately 38% of the bit period); the number of connectable terminal equipments must not exceed 4.

Other configurations can be implemented with the aid of appropriate NT functions, e.g. a star configuration or an "active bus" (in which case the sockets for terminal equipments contain storing or processing components).

4.2.2.2 Electrical Characteristics for Information Transmission

All the component signals, i.e. the D-channel signals, B-channel signals, control signals, etc. (cf. Sect. 4.2.2.4) are combined by time-division multiplexing; a multiplexed signal is thus transmitted in each direction on a wire pair (cf. Sect. 4.2.2.1). Coupling is effected via transformers or their equivalents.

In the case of the passive bus, it is possible that some or all of the terminal equipments (on the same wire pair) may be transmitting at the same time. Normally this only applies to framing signals (see Sect. 4.2.2.4) and D-channel

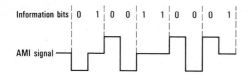

Information bits | 0 | 1 | 0 | 0 | 1 | 1 | 0 | 0 | 0 | 1 |

AMI signal

Fig. 4.9. Transmission Code for the Basic Access: AMI Code with 100% Pulse Width.
One-bit: no pulse (space)
Zero-bit: Pulse (successive pulses have opposite polarity)

signals (see Sect. 4.2.2.3). The procedure for access of terminal equipments to the D channel (see Sect. 4.2.2.3) requires that the NT always receives a "zero" bit when at least one terminal equipment is transmitting a "zero" in the D channel, and that the NT only receives a "one" when all the terminal equipments are transmitting a "one" or nothing. This requirement is met by the following specifications:

- The terminal equipments of a bus configuration transmit bit-synchronously, as all the terminal equipments derive their transmit timing from the same signal received from the NT; the NT for its part derives its timing from the network (see Sect. 4.2.2.1 for delay limitations).
- The transmission code used is AMI [4.16] with 100% pulse width. Unlike the conventional AMI code, a zero is transmitted as a mark (pulse) and a one as a space (no pulse) (Fig. 4.9). This variation was specified to take account of D-channel access by the terminal equipments (see Sect. 4.2.2.3) and of the link access procedure (see Sect. 4.3.4.2) (fill characters between information frames are sequences of ones).
- Suitable specifications for the frame structure (see Sect. 4.2.2.4) ensure that all terminal equipments send zero-bits with the same polarity in the D channel, so that the pulses corresponding to zeros cannot cancel each other out.
- The terminal equipments can transmit by injecting either current or voltage. Even if some or all of the terminal equipments transmit a pulse simultaneously, the voltage at each transmitter output must stay within certain tolerance limits; this is achieved by each transmitter controlling the current or voltage it injects as a function of its output voltage (voltage-limited current or voltage injection). All the transmitter outputs and receiver inputs always present high impedance, even when deactivated; certain limitations apply only during transmission of a pulse.

Table 4.4 shows an overview of the main CCITT-approved electrical characteristics for the basic access (for further details see CCITT Rec. I.430 [4.14]).

4.2.2.3 D-Channel Access by Terminal Equipments

To prevent mutual interference between the terminal equipments in bus configurations during simultaneous transmission in the D channel, these equipments must observe a defined access procedure.

In the case of B channels, normally only one terminal equipment transmits at a time. The exchange ensures this by assigning each of the two B channels to a single terminal equipment at any one time using the signaling.

Table 4.4. Electrical Characteristics for Information Transmission via the Basic Access (in accordance with CCITT Rec. I.430 [4.14]

Characteristics	Values
Characteristics of interchange circuit	
– Terminating resistor	$100\Omega \pm 5\%$
– Maximum attenuation (at 96 kHz) in the point-to-point configuration	6 dB
– Minimum longitudinal conversion loss (at 96 kHz)	43 dB
– Upper limit value for NT-TE-NT round trip delay[a]	42.0 μs
– Maximum NT-TE-NT round trip delay for all terminal equipments in the passive bus arrangement:	
NT without phase control[b]	3.6 μs
NT with phase control[b]	2.6 μs
– Maximum differential round trip delay NT-TE-NT for all terminal equipments in the extended passive bus	2.0 μs
Transmitter characteristics	
– Voltage-limited current or voltage feeding	
– Pulse amplitude at 50 Ω	750 mV $\pm$ 10% = V_{nom}
400 Ω	90[c] to 160% V_{nom}
5.6 Ω	$\leqq$ 20% V_{nom}
– Output impedance	
when transmitting a pulse	$\leqq$ 20 Ω
during a space	Specified values depend on frequency[d]
– Minimum longitudinal conversion loss	54 dB
– Jitter in NT output signal (peak-to peak)	$\leqq$ 0.26 μs
– Jitter in TE output signal	$\leqq \pm$ 0.36 μs
– Phase displacement in the TE between input and output[b]	$-$ 0.36 μs to $+$ 0.78 μs
– Frame offset between TE input and TE output	10.4 μs
Receiver characteristics	
– Input impedance	Specified values depend on frequency[d]
– Minimum longitudinal conversion loss	54 dB
Radiated emissions	Not yet defined
Insulation requirements	Not yet defined

[a] Due to the D channel access procedure (see Sect. 4.2.2.3).
[b] Not taking into account the frame offset of 10.4 μs between TE input and output.
[c] See tolerance mask (in CCITT Rec. I.430 [4.14]).
[d] See CCITT Rec. I.430 [4.14].

Actions Prior to Transmission

A terminal equipment only begins transmitting in the D channel if it has first ascertained by monitoring that the D channel is free in the direction toward the NT. The relevant criterion is at least eight consecutive ones. The link access procedure (see Sect. 4.3.4.2) ensures that this "pause signal" never occurs within a transmitted frame but only between information frames, and that each information frame begins with a zero.

Actions During Transmission

During transmission the terminal equipments checks, by monitoring in the D channel and by comparison, whether the transmitted information is being corrupted by other terminal equipments transmitting at the same time. The electrical specifications (cf. Sect. 4.2.2.2) are such that terminal equipments transmitting a zero win over terminal equipments transmitting a one bit, at the same time. The winning terminal equipment may continue to transmit, while the other terminal equipments must cease transmitting before the next bit. The link access protocol (see Sect. 4.3.4) ensures that the frames from different terminal equipments all have a different content. Thus, after a short time only one terminal equipment will still be sending zero bits. Only this terminal equipment can successfully complete sending its frame while all the other terminal equipments must stand down. As soon as the D channel is free once more (criterion: at least eight consecutive one bits; see above), the terminal equipments that had to yield retry to transmit their information frames.

Priority

The priority for signaling, which is higher than for packet-mode data (cf. Sect. 4.2.1.1), is governed by the number of consecutive "ones" which a terminal equipment must allow to elapse before it can commence transmission; the more ones, the lower the priority class. Terminal equipments which have successfully completed transmission use the same principle to temporarily downgrade their own priority, so as to give way to waiting terminal equipments of the same priority class.

D-Echo Channel

In order to save the expense of providing the terminal equipments with a separate receiver for monitoring in the transmit direction, the NT loops the D-channel information originating from the terminal equipments back to these equipments in a separate *D-echo channel* (cf. Fig. 4.10). The terminal equipments therefore only need to evaluate the D-echo channel for monitoring purposes.

4.2.2.4 Frame Structure

All the control signals and user signals in both transmission directions (Fig. 4.10) are time-division-multiplexed to form a 48-bit frame transmitted 4000 times per second. This corresponds to a total bit rate of 192 kbit/s.

Rapid, unambiguous frame alignment is provided by using a double AMI code violation (see Sect. 4.2.2.2 and example in Fig. 4.10); this is detected no later than the 14th bit in the frame. In the NT direction, all the terminal equipments simultaneously transmit the signals required for frame alignment (F, F_A and the associated L signals). Frame alignment is ensured even if the wires of a wire pair are reversed. An exception applies to bus configurations where the wires for transmission from TE to NT must not be reversed as otherwise the pulses of simultaneously transmitting terminal equipments would cancel each other out.

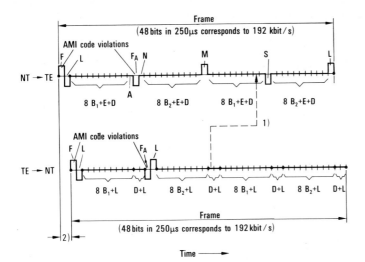

Fig. 4.10. Frame Structure for the Basic Access. To Demonstrate Essential Features, this Example has Binary Ones for all B- and D-channel Bits.

●—●	Frame section with no DC component	N	Inverted value of F_A (therefore usually $N = 1$)
A	Bit used for activation protocol	NT	Network termination
B_1	Bit within B-channel 1	M	Multiframing bit (usually $M = 0$)
B_2	Bit within B-channel 2	S	Bit reserved for future expansions (currently $= 0$)
D	D-channel bit		
E	D-echo-channel bit	TE	Terminal equipment
F	Framing bit	1)	Loopback of a D-channel bit in the D-echo-channel
F_A	Auxiliary framing bit (usually $F_A = 0$; for Q bit option see Rec. I.430)	2)	The terminal equipments transmit with 2 bits frame offset
L	DC balancing bit		

   With the transformer coupling method used (cf. Sect. 4.2.2.2) dc components would cause interference; therefore, the transmitter balances the dc component of each "frame section" delimited by dots in Fig. 4.10 independently by an L pulse of appropriate polarity.

   The first pulse of each frame section is transmitted by the terminal equipments with opposite polarity to that of the framing bit (Fig. 4.10). This ensures both that the pulses for the D-channel signal are transmitted with the same polarity by all the terminal equipments and therefore cannot cancel each other out, and that the AMI violations required for frame alignment (see Sect. 4.2.2.2) occur.

   In order to enable the octet structure essential for PCM-encoded voice transmission to be retained, the signals for the two B channels (B1 and B2) are each assigned as octets to eight successive bit positions in the frame, whereas the signals for the D channel and the D-echo channel occupy non-contiguous bits in the frame because of the D-channel access protocol (see Sect. 4.2.2.3 and CCITT Rec. I.430 [4.14]).

CCITT Rec. I.430 has the option of using every fifth F_A bit (TE to NT) for an auxiliary "Q" channel – possibly subdivided by means of the "M" bit into four subchannels. Similarly, in the opposite direction (NT to TE) an "S" channel can be defined. The S-channel structure provides five sub-channels each of which can transfer one 4-bit character per 5 ms. (The S-channel structuring will be incorporated in the revised version of CCITT Rec. I.430.) The S and Q channel may be used to support additional operation and maintenance functions such as B-channel loopback request or indication and self-test request, indication or report.

4.2.2.5 Activation and Deactivation

The activation and deactivation protocols enable the network to transfer the network termination and the network-fed terminal equipments to an energy-saving status in periods of inactivity. However, it is necessary for these units at all times to be capable of returning to active status, e.g. in the event of an incoming call.

Activation

The trigger for activation (Fig. 4.11) can originate from the exchange (*Activate* signal) or from any terminal equipment (*Initiate activation* signal). All equipments, such as telephones, which can expect incoming calls must therefore understand the *Activate* signal from the exchange even in the idle phase; consequently, during this phase only those functional units that are not required for detection of this signal may be deactivated in these equipments.

The terminal equipments inform the network that their activation is complete using the *Activation completed* signal; this signal must be transmitted by the terminal equipments not later than 100 ms after receipt of the *activate* signal. The network then acknowledges with an *Activation completed* signal.

Deactivation

Deactivation (Fig. 4.12) is only initiated by the exchange (*Deactivate* signal). All the terminal equipments then assume the deactivated status and confirm this

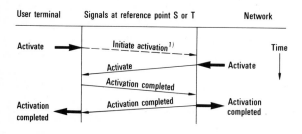

Fig. 4.11. Activation Protocol.
[1]Only if activation is initiated by the user terminal

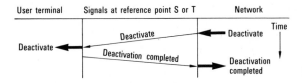

Fig. 4.12. Deactivation Protocol

with the signal *Deactivation completed*. The *Deactivate* and *Deactivation completed* signals are coded in such a way that no pulses – not even the frame alignment pulses – are transmitted. Hence in bus configurations the NT does not receive the *Deactivation completed* signal until all the terminal equipments have been deactivated.

It is left to the network to decide whether and when to deactivate. For example, a network can keep all or specific customer accesses permanently active, or deactive them as soon as all the connections on the access have been released and all signaling activities have ceased.

The NT receives no electrical signal also in cases when the terminal equipments connected to a passive bus are out of service (e.g. unplugged or switched off). It is then up to the network to decide when to deactivate.

4.2.2.6 Electrical Characteristics for Power Feeding

Terminal equipments can. be powered from the NT. There are two methods available for this purpose (Fig. 4.13): the method normally used is to supply power in a phantom circuit via the same four wires on which information is transmitted (Fig. 4.13: source 1/sink 1); alternatively two special additional wires can be used (Fig. 4.13: source 2/sink 2). It depends on the type of NT used whether the NT provides power feeding at all, and if so whether method 1 or 2 is used.

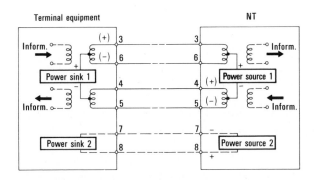

Fig. 4.13. Power Feeding of Terminal Equipments by the NT.
+   –   Feeding voltage during normal operation (reversed in emergency operation)
(+)(−)   Voltage when transmitting a framing bit pulse (Fig. 4.10)
3 . . . 8   Access leads

Table 4.5. Electrical characteristics for Power Feeding of Terminal Equipments at the Basic Access

Characteristics	Power feeding methods (cf. Fig. 4.13)		
	Source 1/sink 1		Source 2/ sink 2
	Emergency condition	Normal condition	
Feeding power available at NT output	$\geq$ 420 mW	[a]	[a]
TE input	$\geq$ 380 mW[c]	[a]	
Feeding voltage: NT output	$-(40\ V + 5\%... - 15\%)$	$+(40(V + 5\%... - 15\%)$	[a]
TE input	$-(40\ V + 5\%... - 20\%)$	$+(40\ V + 5\%... - 40\%)$	
Maximum power consumption per TE when deactivated	25 mW[b]	100 mW	[a]
Current transient	$\leq$ mA/μs		[a]

[a] Depends on network and NT design.
[b] Up to 100 mW may be specified by the network operator for a transitional period.

In the case of phantom power feeding, the NT informs the terminal equipments via the polarity of the feeding voltage (Fig 4.13) whether it is providing only the specified emergency power (emergency operation) or a higher normal power (normal operation). The purpose of emergency operation is to enable users, for instance, to telephone even if the public AC mains supply fails. An NT which is powered during normal operation from the AC mains can be remotely fed by the exchange during emergency operation, and then supplies only selected terminal equipments authorized to receive emergency power.

Table 4.5 gives an overview of the electrical characteristics specified for power feeding.

4.2.3 User-Network Interface for the Primary Rate Access

With this interface the user equipment obtains access to the ISDN at reference points S and T on the basis of the interface structures defined for the primary rate access (see Table 4.3). The associated specifications are given in CCITT Rec. I.431 [4.17].

Two interface variants are defined for the primary rate access, one with a total bit rate of 2048 kbit/s, the other with a rate of 1544 kbit/s. Many countries including the European countries use the 2048-kbit/s variant; the 1544-kbit/s variant is used by the USA, Japan and Canada among others. The two different gross bit rates correspond to the digital transmission systems that are used in the trunk networks of these regions (see Sect. 7.2).

In order to provide easier connection of user equipment to both variants in future, the mechanical characteristics (e.g. connectors) still to be finalized and any new features will require standardization in a uniform manner.

The primary rate access is only available in the point-to-point configuration; there is no passive bus (cf. Sect. 4.2.2.1), which would be difficult to implement due to the bit period being reduced to less than one tenth that of the basic access.

The maximum length of the wiring between the network termination NT and the terminal equipment is limited by the maximum permitted attenuation (6 dB at center frequency, see CCITT Rec. G.703 [4.18]); the resultant range for typical circuits is approximately 150m.

The user equipment is not fed by the NT, so that there is no facility for deactivating the user station and no relevant specification.

As with the basic access, the traffic and control signals are combined in a frame by time-division multiplexing; hence only a single electrical signal is transmitted in each direction.

Electrical Characteristics

The specifications for the electrical characteristics (bit rate, pulse shape, impedance, transmission code) are different for the two interface variants; they have been adopted as they stand from CCITT Rec. G.703 [4.18]. For the 1544-kbit/s variant, CCITT Rec. G.703 [4.18] specifies both AMI and B8ZS as transmission code. The AMI code is simpler but is dependent on the bit sequence: a fairly long sequence of zeros such as may occur in non-voice services can result in synchronization being lost at the receive end, as pulses are only transmitted in response to ones.

In order to be able to handle communication services requiring bit-sequence-independent transmission via the primary rate access in the same way as via the basic access, the standards only allow the B8ZS code to be used on the subscriber line (CCITT Rec. I.431 [4.17]). (Connections routed over trunks via conventional 1.5-Mbit/s transmission systems using AMI code can only be operated on a restricted basis due to the lack of bit-sequence independence within the network, e.g. at 56 kbit/s instead of 64 kbit/s (see Sect. 7.2.3)).

Frame Characteristics

In both variants the specified frame (Fig. 4.14, cf. Sect. 7.2.3) is transmitted 8000 times a second; the frame length is different (256 bits for the 2048-kbit/s variant, 193 bits for the 1544-kbit/s variant). The frame is divided into 31 or 24 8-bit time slots (time channels) preceded by an 8- or 1-bit time slot for control purposes. This gives a net capacity of 1984 or 1536 kbit/s available for the traffic channels and the signaling D-channel.

Part of the transmission capacity of the above-mentioned time slot for control purposes is occupied by the frame alignment signal. This signal is repeated periodically after 2 resp. 24 frames. As any kind of user information can be transmitted in the traffic channels, care must be taken to ensure that the framing does not lock onto an erroneous position; for this purpose, cyclic test data can be derived from the user information and transmitted in the above-mentioned time slots for control purposes (cf. CCITT Rec. G.704 [4.19] and Sect. 7.2.3). A useful side effect of test data of this kind is that it gives the receiver an indication of the bit error ratio on the subscriber line.

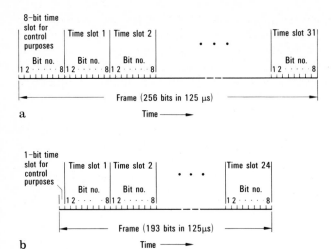

Fig. 4.14a, b. Frame Structure for the Primary Rate Access. **a** Frame structure in both transmission directions for variant with 2048-kbit/s total bit rate; **b** frame structure in both transmission directions for variant with 1544-kbit/s total bit rate

Allocation of 8-Bit Time Slots and Channels

In the case of interface structures incorporating a signaling channel (cf. Table 4.3), in both variants this channel always occupies the same, defined 8-bit time slot: in the 2048-kbit/s variant time slot 16, in the 1544-kbit/s variant time slot 24. If an interface structure without a signaling channel is used, the "signaling channel time slot" is normally reserved in the 2048-kbit/s variant while it is used for a traffic channel in the 1544-kbit/s variant.

In B-channel interface structures (see Table 4.3) each B channel occupies an 8-bit time slot; in H1-interface structures, one H1 channel occupies all 8-bit time slots available for traffic channels.

In order to minimize congestion, in the case of mixed interface structures the standards allow completely flexible allocation of 8-bit time slots to traffic channels (see Table 4.3). The individual 8-bit time slots of an H0 channel need not occur in immediate succession.

Nor do the standards specify any particular time-slot assignment for the H0-interface structures; a possible assignment is described in the Annex to CCITT Rec. I.431 [4.17] (cf. also CCITT Rec. G.735 [4.20] and G.737 [4.21] for sound program applications).

4.3 User Signaling

The main purpose of user signaling is to provide mutual understanding between user terminal and network if services and supplementary services are used (cf.

Sect. 2.3.3) (*user-to-network signaling*). Unlike in conventional networks, the user signaling also allows a degree of transparent information exchange between two user terminals via the D channel (*user-to-user signaling*; see Sect. 4.3.5.5).

Section 4.3.5 describes user signaling for circuit-switched connections. User signaling for packet-switched connections and also for terminal adaptors in circuit-switched and packet-switched connections is described in Sect. 4.4, together with the adaptation of existing interfaces to the ISDN.

User signaling is standardized in a uniform way for all services, for all access types and for all interface structures (cf. Sect. 4.2.1.2) as well as for reference points S and T (cf. Fig. 4.1b). Where a more precise distinction is not required, the following description therefore refers simply to *user equipment* or *user* (TE1, TA, and also NT2 seen from the network) and *network* (NT1, exchange and also NT2 from the perspective of the terminal equipments).

4.3.1 Protocol Architecture

The protocols for user signaling are structured in accordance with the OSI reference model (OSI = Open Systems Interconnection) (see CCITT Recs. I.320 [4.22], I.420 [4.23], I.421 [4.24] and X.200 [4.25], as well as ISO 7498 [4.26]; cf. Sect. 2). In the OSI reference model, functions are abstracted from the communication processes and assigned to seven hierarchical "layers".

User-network signaling takes place in the lower three protocol layers which, broadly speaking, correspond to the following functions (see Fig. 4.15):

The physical layer (layer 1) provides simultaneous, bidirectional transmission of the information signals in synchronism with the network. In the case of the basic access, the physical layer also enables orderly activation and deactivation (cf. Sect. 4.2.2.5) and regulates simultaneous access by several terminal equipments to the common D-channel (cf. Sect. 4.2.2.3).

The data link layer (layer 2) of the D channel serves the network layer by ensuring protected transmission of signaling information and of any packet-mode data transmitted in the D channel in both directions between network and

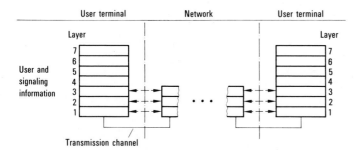

Fig. 4.15. OSI Reference Model for User-Network Signaling and User Information Transfer. Layers 1 to 3: user-network signaling; layer 1: physical layer; layer 2: data link layer; layer 3: network layer

user. The data link layer also allows selective addressing of individual terminals and of groups of terminals in the broadcast case (see Sects. 4.3.3.3 and 4.3.4).

The network layer (layer 3) of the D channel handles user-network signaling in its narrower sense (see Sect. 4.3.5). With packet-switched connections, the network layer, together with the lower layers, is also involved in the transfer of the user information packets (see Sect. 4.4.4).

The protocols of the "higher" layers (layers 4 to 7), also known as application-oriented protocols, are always interpreted by the network users; they may be interpreted, e.g. in the case of messaging services (see Sect. 2.3.1.3), by the network too.

The structuring (not the protocols) of layers one to three, which mainly concern us here, roughly corresponds to the Systems Network Architecture (SNA) protocol structure on which IBM bases its computer networks. However, as regards the actual protocols, the OSI reference model is preferred because it has been further refined – on the basis of the SNA – into an international standard which is neutral in respect of manufacturer [4.25 and 4.26] and is being updated in line with the ISDN (cf. CCITT Rec. I.320 [4.22]).

4.3.2 Types of Connection

The main function of user-network signaling is call control, i.e. control of call establishment and clearing. Three types of connections are distinguished: circuit-switched connections on traffic channels (B, H0, H1), packet-switched connections on traffic channels, and packet-switched connections on the D-channel. In some circumstances the ISDN may also offer a connectionless packet bearer service e.g. for telemetry purposes; this will not be further discussed here.

Circuit-Switched Connections
In the ISDN, user-network signaling for circuit-switched calls (cf. Sect. 4.3.5) only takes place via the separate signaling channel (outslot). This is in contrast to conventional networks such as the analog telephone network, where signaling takes place in the same channel in which the circuit-switched connection is established (inslot). The appropriately modified OSI reference model is shown in Fig. 4.16 (cf. CCITT Rec. I.320 [4.22]). The specifications for the physical layer are given in CCITT Recs. I.430 [4.14] (see Sect. 4.2.2) and I.431 [4.17] (see Sect. 4.2.3); the link access protocol is described in CCITT Recs. Q.920 [4.27] and Q.921 [4.28] (see Sect. 4.3.4), the connection-control procedures in CCITT Recs. Q.930 [4.29] and Q.931 [4.30] (see Sects. 4.3.5 and 4.4).

A circuit-switched connection is only through-connected and handled in the network in the physical layer; the protocols of the other layers are not normally interpreted by the network (Fig. 4.16, upper part).

Outslot signaling has the advantage that signaling (e.g. to indicate a waiting call request) can be accomplished without restriction even if connections already

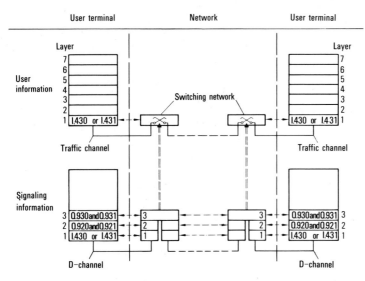

Fig. 4.16. Protocol Architecture for Circuit-Switched Connections

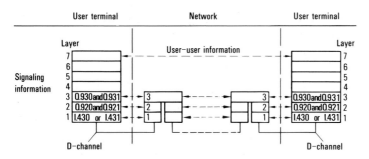

Fig. 4.17. Protocol Architecture for User-User signaling on the D-channel

exist, and that the network can easily assign channels and lines (cf. Sect. 4.2.1.2) to connections and if necessary also assign transmission capacity to channels (cf. Sect. 4.2.3).

However, in case of outslot signaling through-connection of a circuit-switched call is not so closely correlated with signaling as in inslot signaling (cf. CCITT Rec. X.21 [4.3] and Sect. 4.4.2); moreover, here the user equipment has to handle the signaling channel as well as the traffic channels.

User-user signaling can be used to a limited extent by user terminals to exchange packet-mode information via the signaling channel, as on a virtual connection (Fig. 4.17). This information is not interpreted by the network and is transmitted with the original packet sequence retained (cf. Sect. 4.3.5.5).

Packet-Switched Connections

A packet-switched connection on a *traffic channel* (currently only B-channels) is set up in two stages: first circuit-switched connection is between the user equipment and the packet handler (Fig. 4.18) is set up (using outslot signaling); on this circuit-switched connection the conventional packet switching protocols (conforming to CCITT Rec. X.25 [4.4]) are handled (inslot) in the data link and network layers, enabling the virtual connections to be set up and cleared down and the packets to be transmitted (cf. Sect. 4.4.4).

In packet-switched connections on the *D channel* on the other hand, the entire call control process takes place in the same (D) channel as the user information transfer (inslot). As soon as the virtual connection has been set up, layers 1 to 3 handle the transfer of the user information packets (Fig. 4.19 and

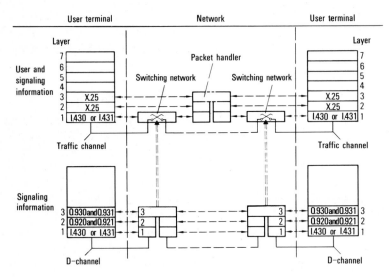

Fig. 4.18. Protocol Architecture for Packet-Switched Connections on a Traffic Channel (e.g. B-channel)

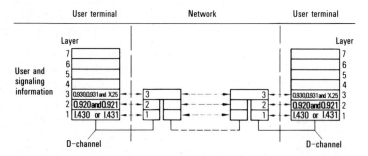

Fig. 4.19. Protocol Architecture for Packet-Switched Connections on the D-channel

Sect. 4.4.4). In addition to conventional packet switching, new modes such as *frame mode bearer service* (cf. Sect. 4.5.1) and *asynchronous transfer mode* (cf. Sect. 4.5.2) can be used.

4.3.3 Special Features of ISDN Signaling

4.3.3.1 Functions of the Network Terminations NT1 and NT2

Whereas the NT1 (cf. Sect. 4.1.2) only performs layer 1 functions, the NT2 (cf. Sect. 4.1.3) may perform a wide variety of functions (Fig. 4.20). It may have no function at all ("zero NT2", Fig. 4.20a and Sect. 4.1.3); in an "active bus" configuration it may handle layer 1 functions only (Fig. 4.20b; the sockets for terminal equipments contain amplification, storage or processing components); as a "statistical multiplexer" it may process layers 1 and 2 (Fig. 4.20c; the information frames of several D-channels are multiplexed in the data link layer); or as a private branch exchange it may process layers 1 to 3 (Fig. 4.20d).

In all four cases (Figs. 4.20a to d) the same set of specifications applies to reference points S and T (cf. Sect. 4.1.1).

4.3.3.2 Call Establishment

In the ISDN, different bearer services and teleservices can be accessed on one line (see Sect. 2.1). At the called end, the ISDN selects not only the correct subscriber line but also a terminal equipment compatible with the required service. To do this, during call establishment the calling terminal gives the network not only address information (ISDN address, see below) but also information on the desired transmission characteristics of the network connection and the compatibility requirements for the called terminal.

Suitable information fields of variable length are provided for this purpose in the setup message (Table 4.6); these contain the bearer capability, ISDN address, compatibility information and user-user information.

Bearer Capability
The bearer capability information element (cf. CCITT Rec. Q.931 [4.30]) contains the transmission characteristics required for the network connection. The main parameters are shown in Table 4.7. Some describe the transmission requirements, e.g. the transmission capacity, the symmetry (bidirectional), the

Table 4.6. Information Fields in the Setup Message

Bearer capability	ISDN address		Compatibility information		User-user information
	ISDN number	ISDN sub-address	for layers 1 to 3	for layers 4 to 7	

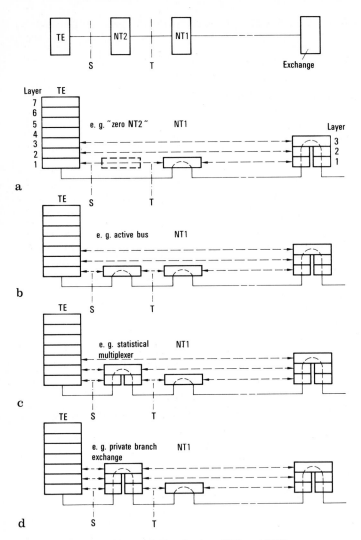

Fig. 4.20a–d. Roles of Network Terminations NT1 and NT2.
NT1, NT2 network terminations, S, T reference points, TE terminal equipment

suitability for specific communication types (e.g. voice or transparent transmission) and the protocol used in layer 1 (e.g. bit rate adaption, cf. Sect. 4.2.1.3 and 4.4). Other parameters describe the mode of establishment of a connection the network connection configuration (e.g. conference) and the switching method (circuit-switched, packet-switched). For packet-switched calls, the protocol to be used in layers 2 and 3 is specified.

On the basis of the given bearer capability, the network checks whether it can provide the desired service (compatibility check in the network). If the test result

Table 4.7. Bearer Capability Parameters (Summary)

Parameter	Possible alternative values
Type of information	– any (bit-transparent transmission (64 or 56 kbit/s)) – speech, audio (3.1- or 7-kHz bandwidth) – motion video – packets
Information transfer mode	– circuit – packet
Information transfer rate	– 64, 2 × 64, 384, 1536 or 1920 kbit/s
Configuration Establishment of connection	– point-to-point – dial-up
Transmission symmetry	– bidirectional symmetric
Protocol options used for transmitting user information	
in layer 1	– standardized speech encoding method (CCITT Recs. G.711, G.721, G722) – bit rate adaption per CCITT recommendations (cf. Sect. 4.2.1.3) – bit rate adaption for HDLC blocks using fillers (flags) – bit rate adaption deviating from CCITT recommendations
in layer 2	– per CCITT Rec. Q.921 [4.28] – per CCITT Rec. X.25 [4.4] (layer 2)
in layer 3	– per CCITT Rec. Q.931 [4.30] – per CCITT Rec. X.25 [4.4] (layer 3)

is negative, the connection request is rejected; if it is positive, connection set-up to the called user equipment proceeds; the latter in turn checks whether it is compatible with the connection request.

Addressing
The ISDN addressing concept goes beyond the "directory number" of the analog telephone network. The ISDN address consists of the ISDN number and optionally the ISDN subaddress (Table 4.8, cf. CCITT Recs. I.330 [4.9] and E.164 [4.31]).

Table 4.8. Structure of the ISDN Address

ISDN address			
ISDN number (max. 15 digits)			ISDN subaddress (max. 20 octets or 40 digits)
Country code	National destination code	Subscriber number	

The ISDN number is used by the network to identify a particular country and the desired access or access group. If appropriate specifications exist between user and network, the ISDN number can be additionally used at the called end for direct dialing-in (e.g. to private branch exchanges), or in some countries for selecting specific terminal equipments in bus configurations.

The ISDN number contains up to 15 digits including the country code and national destination code, but excluding any discriminating digit sequences (e.g. 00 for international calls). The country codes are the same as for the analog telephone network (see CCITT Recs. E.163 [4.32] and E.164 [4.31]).

The national destination code (Table 4.8) can be used to address gateways to dedicated networks or in some cases (e.g. in the USA) to different ISDNs.

The ISDN subaddress is used for more precise addressing of subcomponents within the user premises equipment selected using the ISDN number. It is transferred transparently from the calling to the called user equipment in a specific information field during call establishment. Its length is limited to 20 octets (40 digits).

The ISDN subaddress can also be used by a private network on call establishment in the ISDN to transfer a complete private address to the next (private) node for further network and route selection.

Compatibility Information

While the bearer capability describes the desired transmission characteristics of the connection, the *compatibility information* (see Table 4.6) contains more detailed information on the intended use of such a network connection by the user's equipment. For instance, a terminal adaptor (cf. Sects. 4.1.4 and 4.4) can inform the terminal adaptor on the other side of the network that a transparent 64-kbit/s network connection is to be used with an effective bit rate of 2.4 kbit/s only.

The addressed user equipment only accepts the connection request if it can accept the parameters in the bearer capability and in the compatibility information.

The compatibility information, if any, for layers 1 to 3 (cf. Fig. 4.15) and for layers 4 to 7 is specified in separate information fields; the information field for layers 1 to 3 may contain the same parameters as those defined for the bearer capability (cf. Table 4.7).

The compatibility information relates only to standardized service attributes. The user terminals can also optionally exchange non-standardized information (see *user-user information* field in Table 4.6 and Sect. 4.3.5.5). This information is transmitted transparently by the network and can be used by the terminal equipments e.g. for additional addressing or for compatibility checks.

4.3.3.3 Bus Configurations

The basic access can support bus configurations as well as point-to-point configurations. Bus configurations (see Figs. 4.8b and c) make specific new demands on the signaling. The point-to-point configuration (Fig. 4.8a) is normally handled by the network as a special type of bus configuration.

Unique Terminal Endpoint Identifier

Every terminal equipment operated in a bus configuration must have a different terminal endpoint identifier so that the network can identify the sending terminal equipment when a frame is received and can send an information frame to a specific terminal equipment. The terminal endpoint identifier is incorporated in the transmitted information frames as part of the data link layer address (cf. Sect. 4.3.4).

Knowledge of the Configuration not Required in the Network

As long as there is no connection request outstanding, the network does not need to know the current status of the terminal equipments in a user station, neither their characteristics nor their terminal endpoint identifiers (TEI). When an outgoing call is made, the network knows the TEI of the calling terminal equipment as soon as the latter transmits the connection requests. With incoming calls, the network first transmits the connection request to all the terminal equipments (cf. Sect. 4.3.5). The latter evaluate the connection request by checking the bearer capability and any address and compatibility information contained therein (see Sect. 4.3.3.2). All the terminal equipments for which this check is positive inform the network accordingly. From their response, the network knows the TEIs of the terminal equipments suitable for the connection, and can now selectively address these terminal equipments at least for the duration of this signaling activity.

Depending on the type of additional address information, several, one or even no terminal equipment may respond to an incoming connection request for the service desired. If the connection request contains no additional address information, all the terminal equipments for which the compatibility check is positive respond. If more than one terminal equipment responds, the network assigns the connection to the terminal equipment that responded first and rejects the other terminal equipments (cf. Sect. 4.3.5).

Configuration Flexibility

As the network does not normally recognize the current status of the connected terminal equipments during inactive periods, the user can change the configuration, e.g. unplug, transfer, plug in, switch off or switch on terminals without informing the network. If, however, a connection is to be maintained when a terminal equipment is transferred, the user must first explicitly "park" the connection in the network (see Sect. 4.3.5.4).

4.3.3.4 Simultaneous Signaling Activities

Several circuit-switched connections – one per traffic channel – can exist simultaneously on one subscriber line (cf. Table 4.3). Consequently it has to be possible for several signaling activities – one for each connection – to be handled simultaneously via the signaling channel. Other signaling activities are needed for controlling packet-switched connections (cf. Sect. 4.4) and for registering, controlling or cancelling supplementary services.

In order to enable the transmitted signaling messages to be assigned to the correct signaling activity, every signaling message contains a unique call reference which, in combination with the data link layer address, is unambiguous. The call reference is issued by the entity that initiates the signaling activity, i.e. by the user equipment for an outgoing call and by the network for an incoming call.

4.3.4 Link Access Procedure on the D-Channel

The LAPD (Link Access Procedure on the D-channel) is responsible for protecting all the information (signaling information and packet-mode data, cf. Sect. 4.2.1.1) transmitted on the D-channel in both directions against transmission and sequence errors and for allocating unique terminal endpoint identifiers (TEI, see Sect. 4.3.4.3).

The detailed specifications for LAPD are given in CCITT Recs. Q.920 [4.27] and Q.921 [4.28].

4.3.4.1 Features of the Data Link Layer

The features of the data link layer are oriented towards bus configurations (Fig. 4.8b and c); the point-to-point configuration (Fig. 4.8a) is normally operated by the network as a special type of bus configuration.

The main features offered by the data link layer to the network layer (cf. Fig. 4.15) for information transfer are as follows (Fig. 4.21):

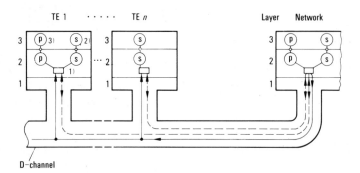

Fig. 4.21. Features of Data Link Layer.
——— Broadcasting of s or p information, – – – – transfer of s or p information from the network to selectively addressed terminal equipments and vice versa.
p packet-mode data, s signaling information, TE terminal equipment.
1) Multiplexing/demultiplexing equipment for s and p information frames, 2) Processing of signaling information, 3) Processing packet-mode data

- Unacknowledged transfer of signaling information (s) and packet-mode data (p) from the network to all terminal equipments (broadcast mode) which process s and p information (continuous line in Fig. 4.21).
- Acknowledged and unacknowledged transfer of s and p information from the network to selectively addressed terminal equipments and vice versa (dashed lines in Fig. 4.21).

In the case if acknowledged transfer (see Sect. 4.3.4.2), the data link layer is responsible for detection of transmission errors, error correction and frame sequence monitoring; with acknowledged transfer, it only handles error detection (incorrectly received information frames are ignored by the receiver).

4.3.4.2 Data Transfer Protocol

The structure of all the protocol elements fully meets HDLC standards developed prior to the ISDN (HDLC: High Level Data Link Control procedure, e.g. ISO 3309 [4.33]).

To operate several terminal equipments simultaneously, including terminal equipments handling both signaling information (s) and packet-mode data (p) (see Fig. 4.21), an address field two octets long is normally used. In the octet transmitted first, a distinction is drawn between s and p information and management information (see Sect. 4.3.4.3); the second octet contains the unique terminal endpoint identifier of the terminal involved or a broadcast identifier (cf. Sect. 4.3.3.3). With regard to access to the D-channel, in the case of the basic access (see Sect. 4.2.2.3), contiguous "ones" are used as fillers between data blocks, and in the case of the primary rate access, contiguous "flags" (bit pattern 01111110) or "ones" are used [4.17] (Table 4.9).

The protocol element coding also complies with the HDLC standards (e.g. ISO 4335 [4.34]; cf. also ISO 7809 [4.35] and CCITT Recs. X.25 [4.4], X.75 [4.36], T.70 [4.37]).

Unnumbered information frames (UI frames) are used for unacknowledged transmission (cf. Sect. 4.3.4.1), in compliance with ISO 4335 [4.34].

For acknowledged data transfer (cf. Sect. 4.3.4.1), the standard [4.28], which initially permitted another three variants, now only specifies the multiblock

Table 4.9. Main Parameters of the Data Transfer Protocol

Parameters	Parameter values	
	Basic access	Primary rate access
Interframe time fill	contiguous ones	contiguous flags (0111 1110) or ones
Modulo	128	128
Window size	for s:1; for p : 3	7
Maximum information field length	260 octets	260 octets
Acknowledgement supervision time	1 s	1 s
Frame repetition in the event of fault	up to 3 times	up to 3 times

method with modulo 128. This method is the functional superset of the original three variants and can be implemented at approximately the same cost thanks to modern microelectronics.

Basic features of the data transfer protocol are that the network and the user equipment have equal access (balanced procedure) and that several layer 2 connections e.g. from different terminal equipments can be operated simultaneously – as far as necessary with different parameters. The data transfer procedure is based on the following principle: the receiver acknowledges error-free reception of information frames to the sender and thus implicitly allows further frames to be sent. If an expected acknowledgement is not received within a specific time, the transmission error is corrected by repeating the frame. Sequence errors such as loss or duplication of a frame are detected from the consecutive numbering of the information frames (sequence number) and corrected; the highest sequence number is followed once more by the lowest in each case.

The window size (Table 4.9) determines how many frames the sender may transmit without receiving appropriate acknowledgements: for example, with window size 1 the sender must delay transmission of the next information frame until the acknowledgement for the previous frame has arrived. In order to ensure that the sequence number is unambiguous, the window size must always be smaller than the modulo. The window size, in case of the basic access, is 1 for signaling information and 3 for packet-mode data; for the primary rate access, the window size is generally 7. The maximum information field length is 260 octets in all cases.

The transfer procedure specification is similar to LAPB (Link Access Procedure Balanced, see CCITT Rec. X.25 [4.4]), but somewhat simplified compared with LAPB (see Sect. 4.4).

4.3.4.3 Assignment of Unique Terminal Endpoint Identifiers

As described in Sect. 4.3.3.3, every terminal equipment in a bus configuration must have a different terminal endpoint identifier (TEI). A management entity assigned to the data link layer automatically handles TEI assignment on the network side: as soon as a terminal equipment requires a new TEI value (e.g. after the voltage supply has been restored), it sends a request to the network (Fig. 4.22a) and is then assigned a free TEI value by the network. The network ascertains which TEI values are available by interrogating the terminal equipments (Fig. 4.22b) to check whether a certain TEI value has already been used; if no terminal answers after two interrogations, the TEI value in question is considered to be free.

Normally the network will hold a store of free TEI values so that it can assign a TEI immediately in response to a TEI request, thereby eliminating the time-consuming test indicated in Fig. 4.22a. As well as testing the TEI values individually, the network is also able to test all TEI values simultaneously by sending "test TEI" to all terminal equipments on a broadcast basis, as in Fig. 4.22b (cf. Sect. 4.3.4.1).

User terminal   Protocol elements at reference   Network
                  point S or T

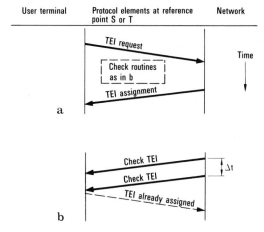

a

b

Fig. 4.22a, b. TEI Assignment Protocol.
TEI Terminal endpoint identifier

The network can also declare on a broadcast basis one or all of the TEI values in the terminal equipments invalid; the network will use this option if it concludes from certain protocol deviations that the same TEI value is erroneously being used by more than one terminal equipment.

The TEI value can also be present in terminal equipments. In this case the terminal equipment must have the TEI value it requires tested for uniqueness by the network according to the same rules (Figs. 4.22a and b). The TEI value to be tested is transferred in the TEI request (Fig. 4.22a).

The TEI assignment protocol elements (Fig. 4.22) are differentiated from the signaling information and packet-mode data by a special management entity identifier in the first address octet (cf. Sect. 4.3.4.2). During TEI assignment, no TEI value is yet available for addressing the terminal equipment concerned; a terminal-generated random number 16 bits long is therefore used for addressing the terminal equipment in this phase.

4.3.5 Signaling for Circuit-Switched Connections

The signaling handled by the call-control procedure (Fig. 4.15) enables the calling user to select the desired user and communication service. The signaling protocol specified for reference points S and T is uniform for all services. The associated standard is contained in CCITT Recommendations Q.930 [4.29] and Q.931 [4.30].

4.3.5.1 Simple Call Establishment

Figure 4.23 shows a simple example of circuit-switched call establishment for telephony. Let two telephones (x and y in Fig. 4.23) be arranged in a bus configuration at the called end (cf. Sect. 4.2.2.1). Let the ISDN address of the called user be transferred by the calling user equipment as a complete block

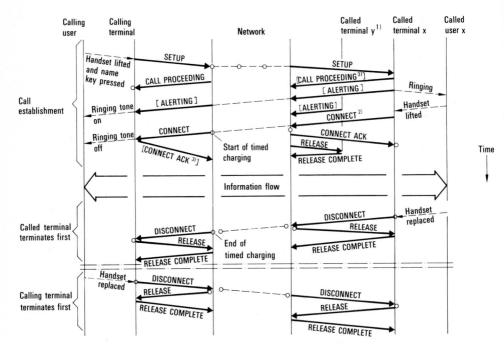

Fig. 4.23. Establishment and Clearing of a Circuit-Switched Call (Set-up with Block Dialing). [. . .] Transfer of this message may be omitted, ○ typical through-connection sequence (for call establishment) and disconnection sequence (for call clearing) for the connection, ACK acknowledge 1) the procedures for the terminal y are not shown; 2) the called terminal x answers before terminal y; 3) permitted for reasons of symmetry (normally ignored by the receiver)

(block dialing); block dialing is a typical feature of user terminals with automatic dialing (e.g. text terminals), but is also frequently used on advanced telephones (e.g. with name key dialing). The user-terminal sequences illustrated in Fig. 4.23 are only intended as examples.

After the name key has been pressed, the calling telephone transfers to the network, in the SETUP message, the bearer capability (see Sect. 4.3.3.2), the ISDN address of the called terminal and the compatibility information for the "telephony" service. The network confirms with CALL PROCEEDING that call establishment is in progress and at the same time assigns a vacant B-channel to the calling user terminal, to which the latter normally switches immediately at latest when receiving the CONNECT message. This mainly applies to telephones in networks providing audible tones (e.g. ringing tone) or announcements in the assigned B-channel at the calling end.

As soon as the destination exchange has detected the connection request, it selects the addressed line, if necessary tests the compatibility and transmits the connection request (SETUP) to all the terminal equipments on a broadcast basis (see Sect. 4.3.4.1). The SETUP message contains the bearer capability and if applicable additional address information and compatibility information (cf.

Sects. 4.3.3.2 and 4.3.3.3). This message is also used to assign a free B-channel: only the specific terminal equipment to which the connection is subsequently assigned will then switch to this B-channel (see below). All the terminal equipments which process signaling information and are compatible with the desired service (terminals x and y in Fig. 4.23) inform the network accordingly with ALERTING (the telephones "ring"), and only send CONNECT when they are able to accept the call (in the case of telephones: "handset lifted"). The connection is now switched through by the network and assigned, on CONNECT ACK, to the terminal equipment which had first sent CONNECT. The other terminals are cleared with RELEASE and confirm this with RELEASE COMPLETE.

The calling terminal is informed of the progress of call establishment by the ALERTING and CONNECT messages: ALERTING corresponds to the ringing tone, CONNECT means that the connection has been switched through within the network. The calling and called user terminals are informed by the signaling in the D-channel when the connection is switched through in the network, but not when it is switched through from user terminal to user terminal (end-to-end) (cf. Sect. 4.4.2).

To facilitate direct connection of private branch exchanges (without the public network), the layer 3 protocol has been defined symmetrically as far as possible; for this reason, sending of the CALL PROCEEDING and CONNECT ACK messages, for instance, (Fig. 4.23) is also permissible for user terminals equipments.

If the connection cannot be set up as desired, the user terminal is notified and informed of the cause.

The signaling information is transferred with acknowledgement by the data link layer as far as possible (see Sect. 4.3.4). Only broadcast messages (e.g. normally only SETUP on the called side; cf. Fig. 4.23) have to be transferred unacknowledged, as the data link layer does not provide acknowledged transfer for broadcast messages.

4.3.5.2 Simple Call Clearing

The calling user terminal, the called user terminal and the network can initiate connection clear-down at any time and independently of each other by sending the DISCONNECT message (lower half of Fig. 4.23).

As discussions stand within CCITT, call clearing is handled as follows:

If the initiative for call clearing comes from a user terminal, the latter disconnects itself from the traffic channel and releases it before sending the DISCONNECT message. After receiving the DISCONNECT message, the network disconnects the traffic channel from the connection within the network, initiates connection cleardown within the network and acknowledges with RELEASE. Finally the user terminal terminates signaling activity (cf. Sect. 4.3.3.4) by sending RELEASE COMPLETE, whereupon the network in turn terminates signaling activity.

On the other, "passive" side of the network, the network likewise disconnects the traffic channel from the internal network connection and, with the DIS-CONNECT message, requests the user terminal to disconnect itself immediately from the traffic channel.

The user terminal confirms this with RELEASE; the network finally releases the traffic channel and terminates signaling activity by sending the RELEASE COMPLETE message, on receipt of which signaling activity is also terminated for the user terminal.

If the initiative for disconnection comes from the network, both user terminals are informed by the network with the DISCONNECT message. For both user terminals, the subsequent course of events corresponds to the procedure shown at the bottom right in Fig. 4.23.

4.3.5.3 Refined Call Establishment and Clearing

Call Establishment with Digit-by-Digit Dialing at the Calling End
Manually operated terminals do not necessarily have to collect the digits of the ISDN number into one block (block dialing), but can transfer the digits to the network singly or in groups.

Direct Dialing in with Point-to-Point Configurations
Part of the ISDN number can be used for direct dialing in to private branch exchanges or comparable installations (cf. Sects. 4.1.5 and 4.3.3.2). The network then transfers to the called user equipment (e.g. PABX) that part of the ISDN number not required for selecting the line either as a block or digit by digit; however, digit-by-digit transfer is only possible in the point-to-point configuration as otherwise unaddressed terminals connected to the bus could compete for the connection prematurely.

Terminal Selection in Bus Configurations
In some countries or networks, part of the ISDN number can be used for selecting specific terminals in bus configurations; for this purpose, the called terminals must know their individual selection address and compare it with the remainder of the ISDN number transferred in the SETUP message (Fig. 4.23).

Channel Negotiation by the Calling User Terminal
For private branch exchanges (PBX) it can be useful in the case of an outgoing call to through-connect the switching matrix before sending SETUP ("forward through-connection"), and for the PBX itself to negotiate a B-channel for the connection. The ISDN permits the calling user terminal to propose the channel but reserves the right to reject the proposed channel.

Channel Negotiation by the Called User Terminal
In packet-switched connections (see Sect. 4.4.4) the called user terminal can inform the network of the channel on which it wishes to receive the connection:

on a new traffic channel or on one already being used by it (currently only a B-channel) or on the D-channel. This channel negotiation facility can also be useful in circuit-switched connections: a called private branch exchange can use it to avoid possible blocking of its switching matrix. For circuit-switched connections, however, this negotiation facility is provided by the network only as an option and not for bus configurations.

4.3.5.4 Control of Supplementary Services

Many supplementary services (cf. Sect. 2.3.3) must be registered in advance in the network before they can be called up or activated; standardized operational procedures are therefore required for registering and handling supplementary services. International standards only exist for a few supplementary services.

In the absence of international specifications, national or network-specific specifications are likely to provide an interim solution.

4.3.5.5 User-User Signaling

The ISDN can optionally provide user-user signaling on the D-channel as a supplementary service. With this facility, user terminals have a limited capability, as with a virtual connection, of sending each other information not interpreted by the network during the establishment (cf. Sect. 4.3.3.2), clear-down and holding time of a circuit-switched connection. Possible applications include exchange of keywords during call establishment, non-standardized communication between two user terminals about the use of an established connection, and above all signaling for tie lines between private branch exchanges.

User-user signaling is a network option and can be organized in a great variety of ways by the individual networks in terms of the maximum permitted information length per packet (35 or 131 octets), throughput and call charges.

There are two possible modes of transfer: piggyback transfer as a part of a signaling message forwarded from user terminal to user terminal, (e.g. SETUP, CONNECT, etc; cf. Fig. 4.23), and transfer with a special message defined for that purpose [4.30].

4.3.5.6 Stimulus Protocol

Review of the Functional Protocol

The handling of terminals (input and output units) and the use of procedures by the terminal user already varies considerably for the terminals employed in conventional networks. Present-day telephones, for example, use different numbers of keys assigned in different ways as input units as well as different visual displays (e.g. a variety of lamp indications and displays in a variety of arrangements) and a wide range of acoustic signal generators as output units. The user-terminal interface with non-voice terminals, e.g. text and data terminals, is usually quite different again. In the use of voice (announcements) and text, the handling of the user-terminal is also dependent on the particular national or regional language.

Further developments will continue to occur in this area: integration of services in the ISDN will favor the trend towards multifunction terminals in which the same input and output units are used for different communication functions and local functions. Further impetus can be expected from progress in sensor technology, voice input and output systems, etc.

In order to avoid burdening the public network with this growing number of possible interfacing solutions, but without letting the public network stand in the way of new solutions, the signaling procedures at the ISDN user-network interface are described as functionally as possible (see Sects. 4.3.5.1 to 4.3.5.5) i.e. as independently as possible from the concrete physical input/output units of a specific terminal.

A software module in the terminal undertakes the conversion between the user-terminal interface procedures (actions of the user and indications for the user) and the functional signaling events at the user-network interface. Hence, the optimum user interface solution can be implemented for each terminal without affecting the network.

Concept and Applicability of the Stimulus Protocol

Although as a general approach the *functional protocol* concept described above is the answer to the requirements discussed, in some circumstances it can also be useful to operate certain terminal types with special functions realized essentially in the network. This applies especially to few but widely distributed terminal types, such as terminals designed exclusively for telephony. The *stimulus protocol*, standardized in outline (see CCITT Rec. Q.931 [4.30] and Q.932 [4.38]) is intended for a user-terminal interface optimized in this way.

The stimulus protocol is based on a hypothetical "stimulus mode terminal", the functional units of which, especially the input and output units, are controlled as directly as possible by the network (Sect. 5.1). It is ensured that stimulus mode terminals can be operated in conjunction with "functional terminals" in the same bus configuration and can also be connected (via the network) to functional terminals.

A stimulus mode terminal reports user actions such as keyboard inputs directly to the network as "stimuli", without itself processing them further. Whereas in the functional protocol it is the terminal that interprets the user actions and converts them to the functional protocol, here it is the network that interprets these stimuli in order to assign them their functional signaling significance depending on previous stimuli or previous states. The stimuli transferred from the network in the other direction are control instructions aimed specifically at functional units in the terminal, e.g. write commands for the screen.

A stimulus mode terminal has the following advantages compared with a functional terminal:

- A stimulus terminal is more flexible in use. By means of suitable control programs in the network, the same terminal can be adapted to suit different applications and different users; new and conventional service attributes can be offered with different user interface procedures and even added subsequently, without preparatory operations or changes (e.g. in the software) being necessary in the terminal. As the logic in the terminal only transmits the stimuli and does not interpret them, a stimulus mode terminal can more easily have a new input or output unit added (e.g. an additional keypad).
- A stimulus mode terminal is – at the network's expense – simpler and thus somewhat cheaper, i.e. more comparable in cost with a single telephone connectable to an analog telephone line.

The stimulus concept has the following disadvantages:

- Greater storage requirements and processing load in the network:
  - The number of transmitted signaling messages is greater.
  - The network must convert between the functional protocol used exclusively within the network and the stimuli transmitted at the user-network interface.
  - The more different terminal types are to be operated, the more program storage must be provided by the network.
  - The network may have to store a large number of variables: terminal-type-specific variables (e.g. tables for the interpretation of key codes) are not critical; line-specific variables (e.g. directory number assignment to name keys) are more costly; terminal- or even terminal-userspecific variables are particularly costly.
- Limited general connectability of the terminals: the terminals can only be usefully operated on another access if the network provides the necessary control software and variables for the new access too (especially problematic for connecting a portable stimulus mode terminal to different networks).

These disadvantages are less serious for connecting terminals to private branch exchanges than for connecting them to the public network, because a private branch exchange can be more readily adapted to suit individual user requirements, and because a terminal normally remains within the area of the private branch exchange.

Current standards define only a general framework for the stimulus concept. Realistic support of a stimulus mode terminal by the network requires additional specifications, including a more precise definition of the characteristics (e.g. the screen capacity) of each stimulus mode terminal, as perceived by the exchange, of the operating procedures.

As seen by the network, a stimulus mode terminal behaves like a video display terminal: the pressing of alphanumeric and function keys by the human operator is conveyed to the network via suitable information elements (KEYPAD or FEATURE ACTIVATION); for its part, the network can reach the

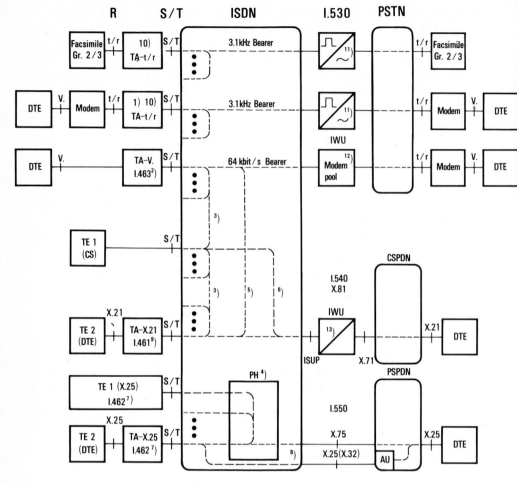

Fig. 4.24. Configuration for the Support of Terminal Equipments with Non-ISDN Interfaces by an ISDN.

[1] Adaption of the analog tip/ring interface: digitized modem signals (indirect adaption of V to S interface, decentralized conversion);

[2] Method in accordance with CCITT Rec. I.463: digital signals (direct adaptation of V to S interface, central conversion in IWU: modem pool);

[3] Only possible for 64 kbit/s (user class of service 30 as defined in CCITT Rec. X.1);

[4] Packet-mode bearer service supported within ISDN by a remote X.25 packet handler (PH, cf. Sect. 4.4.4.1);

[5] Possible for identical user bit rates as a result of commonality in inslot frame structure and bit rate adaption for TA-V. and TA-X.21 (cf. Sect. 4.4.2.2);

[6] Bit rate conversion between non-identical data transmission services provided in the IWU by flag insertion/extraction (use of LAP B in the user information channel) and buffering in the IWU (e.g. between user classes of service 4 and 30 in the case of Teletex terminals);

[7] Identical procedures apply to the combination of an X.25 DTE and its TA as well as to a packet-mode TE 1;

[8] ISDN provides transparent access channel to PSPDN;

[9] Adaption of X.21 bis and X.20 bis is also covered by Rec. I.461;

[10] A/D conversion conforming to Rec. G. 711 (A-law, μ-law);

[11] D/A conversion conforming to Rec. G. 711 (A-law, μ-law);

operator via screen (DISPLAY), via functional indications, e.g. indicator lamps (FEATURE INDICATION) and also via audible tones and announcements in the case of telephones. At the start of the operating sequence, suitable identification information can be notified to the network, enabling it to adapt to the requirements of the particular operator or type of terminal. The strength of the stimulus protocol is its flexibility in terms of individual handling of the operating sequences and with respect to future expansions (without modifying the terminal equipment). This flexibility is based on the possibility of direct information exchange between user and network. This strength is mainly apparent in the individual handling of supplementary services or in the subsequent introduction of new service attributes, rather than in the handling of basic functions such as call setup and cleardown.

Automatic terminals derive no advantage from this flexibility and therefore operate better in accordance with the functional protocol.

4.4 Connection of Terminals with Conventional Interfaces to the ISDN

A flexible ISDN implementation strategy must ensure that, for a transitional period at least, not only new ISDN terminals (TE1) connected via the ISDN-specific S interface (see Sects. 4.2 and 4.3), but also existing terminals (TE2) from other networks can be operated at the ISDN user access (see Sects. 2.3.4 and 4.1.4). The main consideration here is to be able to continue using existing terminals and to take advantage of the higher ISDN transmission rate of 64 kbit/s without having to implement a new interface. The options relevant to the connection and interworking of various terminal equipments are illustrated in Fig. 4.24.

ISDN functions to support non-ISDN interfaces and networks can be subdivided into three areas:

Fig. 4.24 (contd.)
[12] Bit rate adaption in accordance with Rec. I.463;
[13] Bit rate adaption in accordance with Rec. I.461 for identical transmission services.

TE1	ISDN terminal equipment with S interface
TE2	Terminal equipment with other interfaces
	(adaption to the S interface with terminal adaptor TA)
DTE	Data terminal equipment
IWU	Interworking unit
CS	Circuit mode
PSTN	Public switched telephone network
CSPDN	Circuit-switched public data network
PSPDN	Packet-switched public data network
– – – –	Cases of interworking
PH	Packet handler
AU	Access unit
t/r	Analog tip/ring interface

Table 4.10. Adaption of Existing Interfaces for Text and Data Communication to ISDN

Switching principle	Network	Interface at reference point R	CCITT Recommendation for adaption and network interworking	Function of terminal adaptor (TA)
Circuit switching	Telephone network (PSTN)	t/r	I.530	– *Signaling conversion* t/r – S by *TA-t/r* (Fig. 4.24) with digitized modem signals for conventional videotex, facsimile gr. 2/3 and data transmission (*indirect* V.-S adaption)
		CCITT Rec. V.24, V.25, V.25 bis	V.110 (I.463)	– *Signaling conversion* V.25 (bis) – S by *TA-V* (Fig. 4.24) for *direct*, purely digital V.-S adaption for data transmission *without* modem at ISDN subscriber access – *Bit rate adaption* (two-stage per I.461) Bit rates from CCITT Rec. V.5 to 64 kbit/s – *Alignment procedure in the user information channel* Status information transfer between the TAs
	Text/data network (CSPDN)	CCITT Rec. X.21, X.21 bis. X.20 bis	X.30 (I.461); X.81	– *Signaling conversion* X.21, X.21bis, X.20bis – S in TA-X.21 (X.21 bis, X.20 bis) (see Fig. 4.26) – *Bit rate adaption* (two-stage, see Fig. 4.27) X.1–64 kbit/s[b] – *Ready for data alignment procedure in the basic channel* Status information transfer between the TAs
Packet switching	Text/data network (PSPDN)	CCITT Rec. X.25	X.31 (I.462)	– *Bit rate adaption* X.1 – 64 kbit/s Alternative 1: flag stuffing[a] Alternative 2: per I. 461

[a] Cf. HDLC data transfer procedure [4.32].
[b] Three-stage method for X.20bis with additional stage performing asynchronous-to-synchronous conversion using the same technique as defined in V.22 for support of X.1 user rates.

- Adaption of the interfaces per CCITT V. and X. Series Recommendations to the requirements of the ISDN S/T interface by means of special *terminal adaptors* TA (see Sect. 4.1.4). In Fig. 4.24 four main categories of terminal adaptors (TA) are shown. These correspond to the basic adaption possibilities that are summarized in Table 4.10 in conjunction with the relevant CCITT Recommendations: *TA-t/r* supporting the analog tip/ring interface (see Sect. 4.4.3.1), *TA-V.* supporting V.24/RS232C or V.35 (see Sect. 4.4.3.2), *TA-X.21* (see Sect. 4.4.2) and *TA-X.25* (see Sect. 4.4.4).
- *Terminal interworking between ISDN terminals (TE1) and other terminals (TE2) via the ISDN* – at least if the net transmission rate and the switching principle are compatible (see cases of interworking in Fig. 4.24).

Table 4.11. CCITT Recommendations on ISDN Interworking

Interworking Case	Recommendation	Contents
General	I.500	Structure of interworking (IW) Recommendations
	I.510	Definitions, general principles, reference configurations
	I.515	Parameter exchange for establishing compatible IW functions (e.g. modem type selection in case of modem pooling)
	I.332	Numbering principles for IW ISDN-dedicated network
ISDN-ISDN	I.520	Configurations and functions for circuit mode and packet mode services
	I.511	Layer 1 internetwork interface
	X.320	Arrangements for data transmission services
ISDN-PSTN	I.530	Configurations and IW functions: – intra-exchange and interexchange IW, – generation of in-band tones and announcements, – signaling conversion, etc.
	I.516	Modem IW arrangements in the case of TA-V (see Sect. 4.4.3.2): – Layer 1 IW functions – mechanisms for selecting modem type
ISDN-CSPDN	X.321 (= I.540)	General arrangements for IW
	X.81	Detailed IW specification includes: – signaling conversion (ISUP/X.71), – bit rate adaption, etc.
ISDN-PSPDN	X.325 (= I.550)	General arrangements for IW

IW	Interworking
PSTN	Public switched telephone network
CSPDN	Circuit-switched public data network
PSPDN	Packet-switched public data network

- *Network interworking functions* ensure that all the terminals connected to the ISDN can reach communication partners both in the telephone network and in public data networks (see Fig. 4.24). CCITT specifications concerning interworking between ISDN and other networks are summarized in Table 4.11.

4.4.1 ISDN Bearer Service and Public Data Network Access Solutions

The basic possibilities for connecting existing text and data terminals to the ISDN are briefly explained below (Fig. 4.25); these possibilities are the use of *ISDN bearer services*, discussed with reference to adapting the X.21 interface [4.3], and the *access to data transmission services that are provided by public data networks (PDN)*. This latter case is restricted to the adaption of the X.25 interface [4.4].

These two approaches differ mainly in terms of where the switching function for the TE2 terminals connected to the ISDN is handled: either in the ISDN (ISDN bearer service method) or, as before, in switching nodes of an independent public data network (PDN). The latter case is known as the *PDN access solution*; here, in contrast to the ISDN bearer service method, the gateway to the PDN is not only used to communicate with data network users but is also required for ISDN internal traffic between the TE2 terminals. The choice between the ISDN bearer service or the PDN access approach basically depends on national conditions, such as the presence of special data networks, the ISDN introduction strategy, etc.

In the case of the *ISDN bearer service solution* (Fig. 4.25a), ISDN exchanges which provide circuit switching can handle signals from TE2 terminals designed

---

Fig. 4.25a, b. Switching Function for Non-ISDN Terminal Equipments connected to the ISDN, provided either entirely within the ISDN or by accessing a Data Network through ISDN. **a** Use of ISDN bearer services; **b** Access to data transmission services provided by PSPDNs.

TE2	Terminal equipment with R interface (complying with interface Recommendation other than I.-Series)
TA	Terminal adaptor
DTE	Data terminal equipment
IWU	Interworking unit
AU	ISDN access unit of the data network
IP	ISDN port
PDN	Public data network
DSE	Data switching exchange
CSPDN	Circuit switched public data network
PSPDN	Packet switched public data network
CCS	Common channel signaling
S	ISDN user-network interface
X.71	Interexchange signaling for synchronous circuit switched data networks [4.45]
ISUP	ISDN interexchange signaling (ISDN User part, cf. Sect. 6.3): Q.761–Q.766

1) Handled transparently in the ISDN. 2) In accordance with Rec. X.32. ——— Connection between two data terminal equipments both connected to the ISDN; - - - - - connection to a terminal equipment connected to the PDN; ⟹ *outslot* signaling: in the D-channel (subscriber side) or in the common signaling channel (ISDN interexchange side); ⟶ *inslot* signaling: in the user information channel

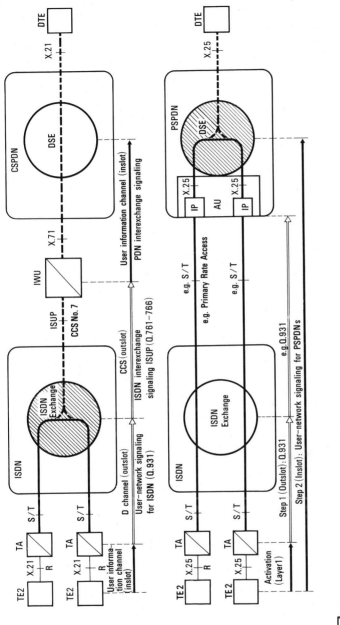

for access to circuit-switched networks; this requires no extension of the functions of the exchanges.

By contrast, connection of terminals for packet-switched networks with X.25 interface [4.4] as part of the ISDN bearer service method, i.e. the provision of ISDN virtual call and permanent virtual circuit bearer services, requires more far-reaching measures in the ISDN (cf. Sect. 4.4.4).

With the ISDN bearer service approach, the functional range of a terminal adaptor (TA) for circuit switching includes the following:

- *Mapping of call control procedures* between the interface at reference point R (e.g. conforming to CCITT Rec. X.21) and the S/T interface (CCITT Rec. Q. 931 [4.30]); the user only needs to carry out *one* call establishment and clearing procedure in accordance with the call control protocol in the data network, e.g. per CCITT Rec. X.21 [4.3]: *single-step call establishment* (see Sect. 4.4.2.1).
- *Bit rate adaption* between the usable data rate of the terminal (cf. CCITT Recs. X.1 [4.39] and V.5 [4.40]) and the ISDN information transfer rate of 64 kbit/s (Sect. 4.4.2.2).
- *Ready for data alignment*, i.e. the task of synchronizing the entry to and exit from the data transfer phase between the two interfaces at reference point R (e.g. per CCITT Rec. X.21), as in data networks. This is achieved with a synchronizing procedure handled in the B-channel (see Sect. 4.4.2.3).

With the *PDN access approach* (Fig. 4.25b) which is confined to packet switching, the only support that an ISDN gives to packet calls is a physical 64 kbit/s circuit-mode transparent connection between an X.25 TA or packet-mode TE1 and the appropriate ISDN port of a packet switched public data network (PSPDN); this method is therefore also known as the *port method*. In this case, although the TE2 terminals are connected physically to the ISDN, logically they remain data network users. This mainly affects subscriber numbering which conforms to CCITT Rec. X.121 [4.41] instead of I.331 [4.31] which applies to ISDN users (I.331 is identical to E.164). In addition, other network functions such as the available service (and possibly supplementary services), call charge accounting, and operation and maintenance functions, are also affected.

With the PDN access solution, data network signaling is not converted to ISDN signaling or vice versa. Instead, call establishment and clearing is always performed in two steps: ISDN user-network signaling (Q. 931) and ISDN interexchange signaling (ISUP) are in this case only used to establish the 64-kbit/s circuit-switched access connection between TA and data network port IP (cf. Fig. 4.25b). On completion of *physical* connection establishment (outslot signaling), another signaling procedure (conforming to the packet layer protocol of CCITT Rec. X.25) is required for setting up the *virtual* circuit in the PSPDN; as in data networks, this procedure is handled directly between TE2 and data exchanges: *two-step call establishment*. In this case X.25 call control procedures (cf. Sect. 4.4.4) are conveyed – transparently for the ISDN – via the ISDN B-channel connection already established (inslot signaling).

4.4.2 Connection of X.21 Terminal Equipment
with Single-Step Call Establishment

The functions (mentioned in Sect. 4.4.1) of a terminal adaptor (TA) designed on the single-step call establishment principle for X.21 adaption in compliance with CCITT Rec. X.30 (identical to I.461) [4.6] are described in greater detail below:

- Mapping of X.21 call control procedure to the D-channel protocol (see Sect. 4.4.2.1)
- Bit rate adaption from the X.1 bit rate to 64 kbit/s (see Sect. 4.4.2.2)
- Ready for data alignment in the 64 kbit/s channel (see Sect. 4.4.2.3).

4.4.2.1 Mapping of the Call Establishment and Clearing Procedures Between the X.21 and S Interfaces

Technical details for mapping the X.21 signaling events to the corresponding D-channel signaling messages (cf. CCITT Rec. Q.931 in Sect. 4.3.5) are given in Fig. 4.26. Only the following remarks need therefore be made here:

- On connection setup the individual dialed digits from the X.21 side are first collected in the TA before a call establishment message (SETUP) is sent with enbloc dialing to the network.
- On cleardown, the DTE Clear Request per CCITT Rec. X.21 causes the TA to initiate the release procedure on the network side both in the D-channel and in the B-channel (see Sect. 4.4.2.3).

4.4.2.2 Adaption Between the X.1 User Rates of X.21 Terminal Equipment and the ISDN Information Transfer Rate of 64 kbit/s

The purpose of the bit rate adaption functions carried out by the X.21 terminal adaptor is to match the X.1 user rate of the connected X.21 terminal equipment (Table 4.12) to the B-channel rate of 64 kbit/s by adding additional and filler information.

With the exception of 48 kbit/s, a *two-stage method* (Fig. 4.27) is specified for this purpose in CCITT Rec. X.30 [4.6]:

- In *stage 1 (RA1)* an intermediate bit rate of 8 or 16 kbit/s is initially produced (Table 4.12) by creating a 40-bit frame in the terminal adaptor (TA). These 40-bit frames (Table 4.13) are exchanged between the communicating terminal adaptors in the B-channel.

Table 4.12. X.1 User Rates and Intermediate Bit Rates as Specified in CCITT Rec. X.30

X.1 user class of service	X.1 user rate in kbit/s	Intermediate bit rate in kbit/s
3	0.6	8
4	2.4	8
5	4.8	8
6	9.6	16
7	48	64

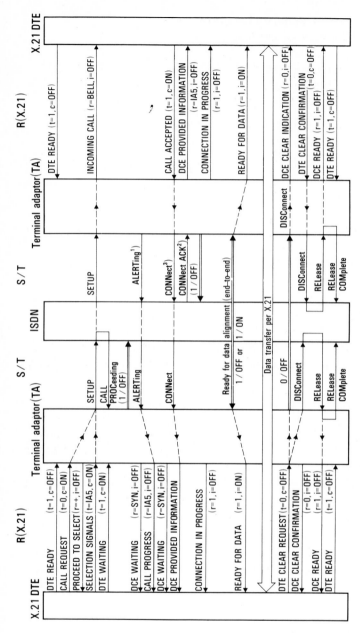

Fig. 4.26. Mapping of X.21 Signaling Events to the Signaling Messages of the D-channel Protocol in the X.21 Terminal Adaptor (TA-X.21).

1) only during manual operation; 2) if several call-compatible TA-X.21s accept an incoming call with ALERTing and/or CONNect, the network sends RELease to the TA-X.21s not selected for the call; 3) if the connected X.21-DTE is in status *controlled not ready*, *uncontrolled not ready or busy*, the called TA-X.21 answers the call with RELease COMPlete instead of ALERTing and/or CONNect. The idle status is reported to the calling side with a DISConnect message.

— D-channel signaling per CCITT Rec. I.451 (cf. Sect. 4.3.5), ═══ signaling in user information channel (cf. Sect. 4.4.2.3)

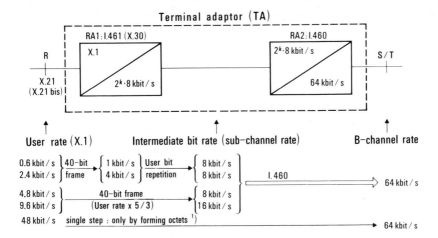

Fig. 4.27. Two-stage Bit Rate Adaptation.
[1] Six user information bits + status bit + frame alignment bit

- In *stage 2* (*RA2*) the intermediate bit rate is increased to 64 kbit/s in accordance with CCITT Rec. I.460 [4.12] by setting all unused bit positions within a B-channel octet to binary "1"; in the 8 kbit/s case, this produces a 1111 111X bit sequence in the B-channel and at 16 kbit/s a 1111 11XX bit sequence.

This relatively complex two-stage method of padding out a *single* bit stream to 64 kbit/s was introduced for reasons of compatibility with *synchronous time-division multiplexing of several independent 8, 16 or 32 kbit/s bit streams*, likewise defined in CCITT Rec. I.460; this allows any combination of these subrate streams (sub-channels) in *one* B-channel (cf. the example in Fig. 4.28). Subdivision of the ISDN user access into sub-channels is initially only planned as part of an end-to-end 64 kbit/s connection between two users.

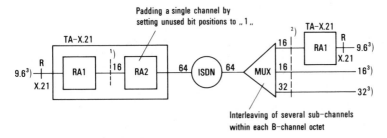

Fig. 4.28. Rate adaption of a Single Bit Stream and Multiplexing of Several Independent Substreams into a 64 kbit/s-channel. All bit rates in kbit/s.
[1] Intermediate bit rate;   [2] sub-channel bit rate;   [3] user bit rate

Table 4.13. Structure of the 40-bit Frame for Bit Rate Adaption in Stage 1 (RA1)

	Bit no.								
	0	1	2	3	4	5	6	7	
Octet no.									
0	0	0	0	0	0	0	0	0	odd frames
	1	E1	E2	E3	E4	E5	E6	E7	even frames
1	1	P1	P2	P3	P4	P5	P6	SQ	
2	1	P7	P8	Q1	Q2	Q3	Q4	×	
3	1	Q5	Q6	Q7	Q8	R1	R2	SR	
4	1	R3	R4	R5	R6	R7	R8	SP	

P, Q, R	8-bit groups for user information
SP, SQ, SR	Status bits
E1 . . . E7	These bits provide additional inslot signaling capacity between TAs, e.g. for inslot coding of the X.1 user rate
×	Used for end-to-end flow control between TAs supporting asynchronous DTEs operating at different user rates

4.4.2.3 Ready for Data Alignment in the 64 kbit/s Channel Between the Terminal Adaptors and Between the X.21 Terminal Equipments

Using the 40-bit frame illustrated in Table 4.13, the terminal adaptors (TA) exchange not only user information (cf. the 8-bit groups for user information P, Q and R) but also the status information provided at the X.21 interface for call control. With the aid of the status bits SQ, SR and SP, the X.21 signaling states on interface circuits t, c, r, i can be transmitted via the ISDN 64 kbit/s connection already set up between the terminal adaptors.

The X.21 interface can also be used at the ISDN with a user bit rate of 64 kbit/s corresponding to user class of service 30 (CCITT Rec. X.1). In this case, because the B-channel rate of 64 kbit/s is fully utilized for data transmission, no status information can be transferred over the 64 kbit/s channel as soon as an X.21 connection has been established. Consequently, it is not possible to use an inslot alignment method at layer 1 when leaving the data transfer phase. For interworking with directly connected ISDN terminals with S interface (TE1), the alignment procedure is also omitted when entering the data transfer phase, as ISDN terminals (TE1) perform alignment above layer 1.

4.4.3 Connection of Data Terminal Equipment with V.-Series Type Interfaces to the ISDN

Non-voice terminals with interfaces conforming to the CCITT V.-Series Recommendations for use in analog telephone networks can be adapted to the S/T interface of the ISDN user access via two fundamentally different terminal adaptors (TA) (see Table 4.10 and Fig. 4.24). Using the ISDN bearer service solution, the S/T interface can either be converted

- to the analog tip/ring interface of the telephone network (TA-t/r), or
- to a V.-series interface for data transmission in the telephone network (TA-V).

4.4.3.1 Support of the Analog Tip/Ring Interface in the ISDN

The terminal adaptor TA-t/r performs analog/digital conversion of the analog (modem) signals of the t/r interface, representing the inverse of the digital/analog conversion at the boundary between the ISDN and the analog telephone network. Besides, decadic pulsing or touch tone signaling is converted to the D-channel protocol.

The main advantage of this method is that the network interworking capabilities for the conventional telephone network required for voice communication can also be used without modification for data traffic.

4.4.3.2 Support of V.-Series Interfaces in the ISDN

The terminal adaptor TA-V specified in CCITT Rec. V.110 (identical to I.463) [4.8] performs a direct conversion of the V.24 interface signals [4.42] to the S interface using purely digital signals. Therefore, no modem is required for internal ISDN traffic. TA-V functions comprise alternate voice/data circuit switched services in conjunction with manual call control performed by a separate digital telephone set.

In addition to that, mapping functions necessary to convert automatic calling and/or automatic answering procedures of CCITT Recs. V.25 [4.43] and V.25bis [4.44] to the ISDN D-channel protocol (CCITT Rec. I.451) are also planned.

As in the case of the TA-X.21, TA-V performs

- Conversion of the electrical, mechanical, functional and procedural characteristics of the V.-series type interface to those of the ISDN S/T interface.
- Bit rate adaption of V.-series user data to 64 kbit/s. By adopting for the TA-V the 40-bit frame and the two-stage rate adaption method defined for TA-X.21 (see Sect. 4.4.2.2), terminal interworking between X.21/X.21bis/X.20bis DTEs and V.-series DTEs via the ISDN is possible (see Fig. 4.24). In contrast to synchronous V.5 bit rates, an additional asynchronous-to-synchronous conversion stage RA0 is added in the case of asynchronous bit rates (three-stage method). In RA0, incoming asynchronous data is padded by the addition of stop elements to fit the nearest channel rate defined by 2^n times 600 bit/s.
- End-to-end alignment of entry to and exit from the data transfer phase. For compatibility with X.30 (I.461), channel status information associated with V.24 interface circuits 108, 107, 105 and 109 is conveyed in-slot by making use of the status bits in the 40-bit frame (see Table 4.13).

A flow control option allows the connection of asynchronous DTEs operating at different user data rates by reducing the output rate of the faster to that of the slower. Between two communicating TAs the x-bit is used to carry flow control information.

4.4.4 Connection of Terminals with X.25 Interface to the ISDN

4.4.4.1 Basic Characteristics

The integration of packet switching into the ISDN [4.45] basically follows the two approaches presented in Sect. 4.4.1:

● With the *ISDN virtual circuit bearer service*, the packet-switching function is part of the ISDN. Virtual connections between packet terminals connected to the ISDN are therefore handled via ISDN-internal X.25 switching devices known as *packet handlers* (PH) or packet switch service modules (PSSM). The PH functions can either be incorporated in an ISDN exchange or located remotely from the ISDN circuit-switching exchanges and assigned as centralized packet switching devices to the higher ISDN network level (Fig. 4.29).

The packet handlers can be fully interconnected via permanent 64-kbit/s ISDN connections, by means of packet-switching interexchange call control and data transfer procedures in accordance with CCITT Rec. X.75 [4.36], thereby forming an overlay network for packet switching within the ISDN. A packet handler can also assume, if required, the function of a gateway to an independent packet network (PSPDN) on the basis of CCITT Rec. X.75.

Fig. 4.29. Provision of ISDN Virtual Circuit Bearer Services by Access to Packet Handling Function within ISDN.

TE1	ISDN terminal equipment with ISDN S interface
TE2	Terminal equipment with non-ISDN R interface
DTE	Data terminal equipment
TA-X.25 (B)	X.25 terminal adaptor for B-channel access to PH, i.e. X.25 link and packet layer procedures are conveyed through the B-channel
TA-X.25 (D)	X.25 terminal adaptor for D-channel access to PH, i.e. X.25 packet layer procedures are conveyed through the D-channel
NT1	Network termination
PH	Packed handler within ISDN
PSPDN	Packed switched public data network
ISUP	ISDN interexchange signaling (cf. Sect. 6.3)
CCS	Common channel signaling
X.25	User interface for public packet networks [4.4]
X.75	Interexchange interface between public packet networks [4.36]
s	Signaling information (SAPI = "s")
p	Packet data in the D-channel (SAPI = "p")
d (PS)	Packet data in the B-channel
SAPI	Service access point identifier (cf. Sect. 4.3.4)

[1] Separation of signaling information from packet data; [2] layer 2 message interleaving of packet data of different ISDN subscriber accesses.

══ Access to the PH via the B-channel: in the net work, an individual connection (i.e. one per TE2) is used (demand or permanent), ── Access to the PH via the D-channel: in the network a permanent connection is used jointly with other TE2 terminals by message interleaving

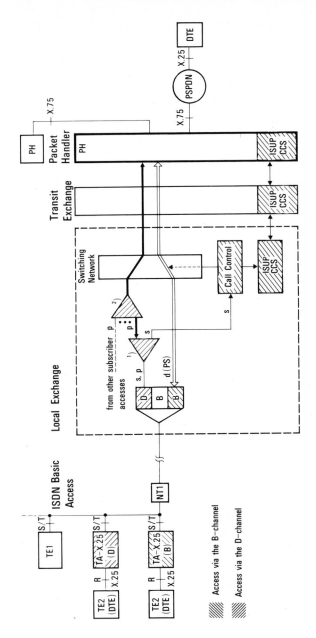

● With the circuit-switched *access to PSPDN services* the packet-switching function of an independent packet network (PSPDN) is used. Virtual connections between ISDN users are thus always routed via a separate PSPDN in which the processing of the X.25 call is carried out (Fig. 4.25b).

In this case, packet-switched services and supplementary services in accordance with specifications for PSPDNs are available to the user (see CCITT Recs. X.1 [4.39], X.2 [4.46] and X.25 [4.4]). In addition, the general procedures and principles of CCITT Rec. X.32 [4.47] apply which define the access to a PSPDN through a circuit-switched public network.

The ISDN procedures for the packet-switching service are given in CCITT Rec. X.31 (identical to I.462) [4.7]. These specifications apply not only to existing X.25 terminal equipments TE2 [4.4], matched to the ISDN user access via a terminal adaptor (TA), but also to future directly-connected ISDN packet-mode terminals TE1 with S/T interface.

Unlike the ISDN bearer service method of connecting terminal equipments designed for circuit-switched networks (Fig. 4.25a), the ISDN virtual circuit bearer service approach is based on the principle of *two-step call establishment* (see Sect. 4.4.1). The two-step procedure for access to PSPDN services reproduced in Fig. 4.25b therefore also applies in principle to access via the B-channel to a packet handler (PH) in the ISDN; cf. the protocol architecture for X.25 based packet mode services in ISDN, reproduced in Fig. 4.30. However, the other characteristics such as subscriber numbering in accordance with the ISDN numbering plan (CCITT Rec. E.164 [4.31]), ISDN virtual circuit bearer services in accordance with CCITT Rec. I.232 [4.48], subscriber administration, etc. correspond to the ISDN bearer service solution described in Sect. 4.4.1.

In the meantime, X.25 packet switching has been supplemented by new packet switching techniques using outslot signaling with single-step call establishment based on a modified Q.931 protocol (see Sect. 4.5).

With the ISDN virtual circuit bearer service method (Fig. 4.29), a packet terminal at the ISDN basic access ($B_{64} + B_{64} + D_{16}$) can either use a *B-channel* (see Sect. 4.4.4.3) or the *D-channel* (see Sect. 4.4.4.4) for end-to-end exchange of X.25 control and data packets with a packet handler. However, the network operator decides which of the access types, described in more detail in the following sections, are actually provided:

– B-channel access only (B)
– D-channel access only (D)
– both access types (B/D).

In networks which have both access types available, the type of channel actually used for the delivery of a new X.25 incoming call packet at the user access can be preset by prior agreement, e.g. in the form of customer profile information stored in the network.

However, if both access types are permitted on one ISDN access, the network must perform channel selection individually for each incoming virtual

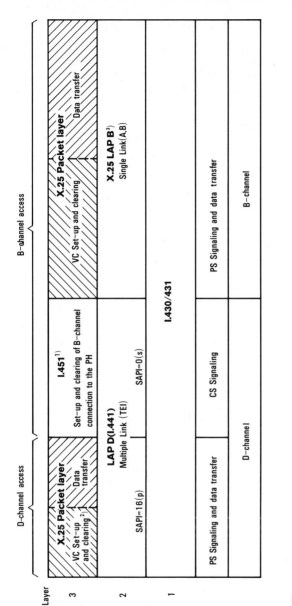

Fig. 4.30. Protocol Architecture for X.25 Based Packet Mode Services in ISDN According to CCITT Rec. I.462 (X.31).

[1] Two-step call establishment: I.451 and X.25
[2] At the called side: additional call offering procedure based on I.451 (cf. Sect. 4.4.2)
[3] I.441 (LAP D) in the case of logical link multiplexing in the B-channel (cf. Fig. 4.36)

SAPI	Service Access Point Identifier	CS	Circuit switched
TEI	Terminal Endpoint Identifier	PS	Packet switched
PH	Packet Handler	VC	Virtual call

call as part of the call establishment procedure, because the ISDN does not have permanent knowledge of the current user configuration (cf. Sect. 4.3.3.3); therefore, in the case under consideration the network does not know whether terminal adaptors (TA) for B- and/or D-channel access are connected to the ISDN bus at the called side; such an adaptor may even be a combined device designed for both access types. Therefore, in the most general case, the network offers every incoming virtual call, i.e. every new X.25 incoming call packet, globally to all packet-mode TAs/TE1s at the called ISDN access by means of a special point-to-multipoint signaling procedure for packet switching known as the *call offering procedure* (cf. Sect. 4.3.5.1).

4.4.4.2 Point-to-Multipoint Signaling for Incoming Virtual Calls

With the ISDN virtual circuit bearer service method a special point-to-multipoint procedure based on the incoming call procedure for circuit-switched calls is used to offer every new incoming virtual call to all packet mode terminals present at the called ISDN access, and to determine the channel type to be used for the delivery of the X.25 incoming call packet: the *call offering procedure* already mentioned in Sect. 4.4.4.1 (Fig. 4.31). As in the case of circuit-switched call control the network selects the first user equipment which responds to the call offering message SETUP with a CONNect message. By virtue of the access possibilities for packet switching offered by the network in the SETUP message (B or D or B/D), the responding call-compatible packet-mode terminals can request, on an individual-call basis by means of the CONNect message, the channel type to be used for each incoming virtual call:

– a new (idle) B-channel to be set up for the virtual connection (*new B*) or
– a B-channel already used by the same packet-mode terminal for other virtual connections to the PH (*established B*) or
– the D-channel (*D*).

   In the event of an *established B* or *D* channel request, the network terminates by means of RELease the signaling activity (signaling transaction) towards the packet-mode terminal selected for the virtual call, as this activity is no longer needed for the subsequent course of the virtual X.25 connection. If, however, the selected packet-mode terminal requests channel type *new B*, the s-type call offering procedure is continued as a normal incoming call procedure (as in circuit switching, cf. Sect. 4.3.5.1) with CONNect ACKnowledge (see B-channel access in Fig. 4.31a). If channel type *D* is requested, the network uses the terminal endpoint identifier (TEI) specified in the CONNect message to establish an HDLC-LAPD data link with $SAPI = p$ (p link) to the packet terminal selected in the preceding call offering phase, provided such a link does not already exist for handling other virtual X.25 connections to the same terminal (see D-channel access in Fig. 4.32a).

4.4.4.3 Access to Packet Switching via the B-Channel

For access to a packet handler in the ISDN (ISDN virtual circuit bearer service) or to an independent packet network PSPDN (circuit-switched access to PSPDN services), a permanent or "on demand" ISDN B-channel connection is used as the transparent feeder between TA-X.25 on the one hand and the packet handler or packet network on the other. As mentioned in Sect. 4.4.4.1, connections are set up in accordance with the two-step call establishment principle (Fig. 4.31); cf. the protocol architecture for X.25 based packet mode services in Fig. 4.30.

Except where a permanent B-channel connection is used or a virtual connection to the same packet terminal already exists, the TA-X.25 for an outgoing virtual connection must first employ the D-channel signaling procedures defined for circuit-switched calls (see Sect. 4.3) to create a 64-kbit/s connection to an input port of the packet handler or packet network: *step 1*; the X.25 terminal can cause the terminal adaptor to do this by means of appropriate activation measures in layer 1 of the X.25 interface, such as the X.21/X.21 bis hotline procedure. When the B-channel connection to the packet handler or packet network is set up, the TA-X.25 puts layer 1 of the X.25 interface into the data transfer state. The X.25 protocols of layer 2 (establishment of the HDLC-LAPB data link) and 3 (virtual call) can then be handled directly between the X.25 terminal equipment and the packet handler or packet network: *step 2*.

At the *called end*, two-step call establishment basically operates in the same way, except that the initiative comes from the packet handler or packet network. The first step covers the multipoint signaling for incoming virtual calls. If Packet Mode (PM) is specified, only packet mode terminals respond to the SETUP message. Finally – as previously at the calling end – the X.25 protocols in layers 2 and 3 are performed directly between the packet handler and X.25 terminal: *step 2*.

CCITT Rec. I.462 does not specify the procedure between the ISDN local exchange and the PSPDN or the packet handler. In Fig. 4.32 it is assumed that the D-channel protocol is employed for switched B-channel connections.

The *call clearing procedure* (Fig. 4.31b) also takes place in two steps in reverse order: release of the virtual connection per CCITT Rec. X.25 is followed – if it is the last virtual connection to the terminal in question – by cleardown of the B-channel connection on the basis of D-channel signaling procedures defined for circuit-switched calls (CCITT Rec. Q.931).

4.4.4.4 Access to Packet Switching via the Signaling Channel (D-Channel)

For access to an ISDN packet handler via the D-channel (Fig. 4.32), packet data *p* and signaling information *s* share the physical transmission capacity of the D-channel in a multiplex arrangement. This is effected in layer 2 on a *message interleaving* basis; distinction is made in the HDLC address field by means of the service access point identifier $SAPI = p$ or $SAPI = s$ (cf. Sect. 4.3.4).

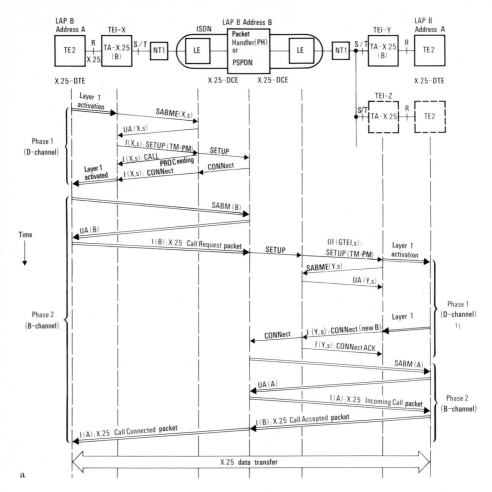

Fig. 4.31a, b. Procedure for Access to Packet Switching Services via the B-channel: ISDN provided Packet-Mode Bearer Service and Access to PSPDN provided Data Transmission Service.
a Set-up of the first virtual call; **b** clearing of the last virtual call.

TE2	Terminal equipment with non-ISDN interface
TA	Terminal adaptor
GTEI	Global TEI (unacknowledged broadcasting to all terminal equipments; cf. Sect. 4.3.4)
TEI	Terminal endpoint identifier (cf. Sect. 4.3.4)
TM	Transfer mode
PM	Packet mode
LE	ISDN local exchange
NT1	Network termination
PSPDN	Packet switched public data network
SAPI	Service access point identifier (cf. Sect. 4.3.4)
———	D-channel signaling (outslot)
═══	Signaling in the B-channel (inslot)

SETUP, ALERTing . . .: ISDN I.451 call control messages (cf. Sect. 4.3.5)
SABM, DISC, UA, UI, I: HDLC commands or responses
[1] Call offering procedure using LAPD with SAPI = "s" (see Sect. 4.4.4.2);
[2] after clearing of the last virtual call

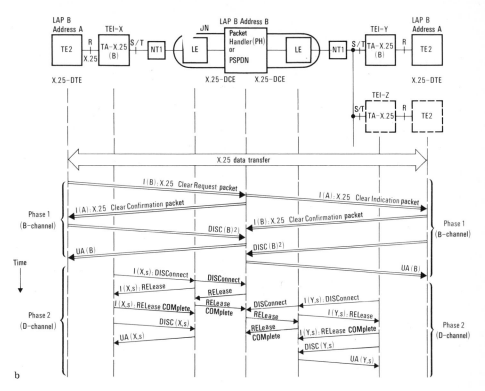

Fig. 4.31b. *Notation of D-channel frames:*
I(Y,s): ALERTing (new B)

- I.451 message (layer 3)
- SAPI
- TEI
- HDLC command (layer 2)

Unlike access via the B-channel, in this case the X.25 packets in layer 2 are only transported across the X.25 interface using the LAPB as defined in CCITT Rec. X.25; between TA-X.25 and the local exchange or packet handler, the D-channel LAPD is used. In comparison with B-channel access, utilization of the D-channel gives rise to certain restrictions in the use of the X.25 protocol, because exchange of signaling information must not be hindered: these restrictions include curtailing the length of the data field in the X.25 data packets to a maximum of 256 octets and limiting the possible packet throughput.

As the common channel signaling network (cf. Sect. 6.3) was designed for the exchange of signaling information between exchanges, in most networks it is not suitable for packet-switched data traffic. Consequently even with D-channel access on the subscriber line, permanent 64 kbit/s channels are used for the further transport of X.25 packets to the packet handler. For this purpose packet data $(SAPI = p)$ in layer 2 of the D-channel must be separated from the signaling information $(SAPI = s)$ in the ISDN local exchange (see Fig. 4.29). The

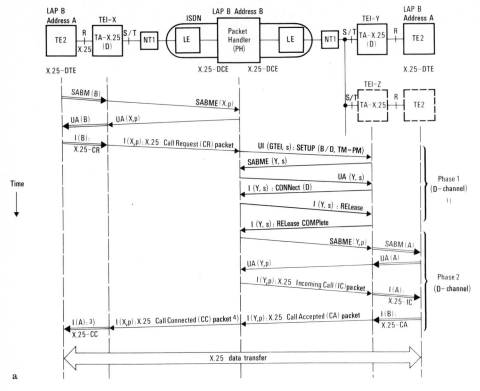

Fig. 4.32a, b. Procedure for Access to Packet Switching Services via the D-channel: ISDN provided Packet-Mode Bearer Service.

a Set-up of the first virtual call; **b** clearing down of the last virtual call

TE2   Terminal equipment with non-ISDN interface
TA   Terminal adaptor
GTEI   Global TEI (unacknowledged broadcasting to all terminal equipments)
TEI   Terminal endpoint identifier (cf. Sect. 4.3.4)
SAPI   Service access point identifier (cf. Sect. 4.3.4)
TM   Transfer mode
PM   Packet mode
LE   ISDN local exchange
NT1   Network termination
SETUP, CONNect. . .: ISDN I.451 call control messages (cf. Sect. 4.3.5)
SABM, DISC, UA, UI, I: HDLC commands or responses
————— D-channel signaling (outslot)
========= inslot signaling

[1] Call offering procedure using LAPD with SAPI = "s" (see Sect. 4.4.2);
[2] after clearing of the last virtual connection;
[3] transport of X.25 packets in layer 2 of the X.25 interface: X.25 LAPB;
[4] transport of X.25 packets in layer 2 of the D-channel: LAPD with SAPI = "p" (LAPB and LAPD data links coupled in the TA)

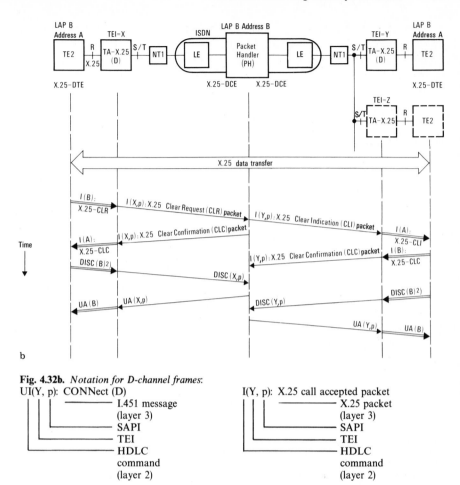

Fig. 4.32b. *Notation for D-channel frames*:

UI(Y, p): CONNect (D)

```
  |  |  |  ┌──────── I.451 message
  |  |  |  |          (layer 3)
  |  |  └──────────── SAPI
  |  └─────────────── TEI
  └────────────────── HDLC
                      command
                      (layer 2)
```

I(Y, p): X.25 call accepted packet

```
  |  |  |  ┌──────── X.25 packet
  |  |  |  |          (layer 3)
  |  |  └──────────── SAPI
  |  └─────────────── TEI
  └────────────────── HDLC
                      command
                      (layer 2)
```

p information is then forwarded on to the packet handler – transparently, i.e. without any processing in the ISDN exchanges through which it passes – e.g. via permanent 64 kbit/s channel connections on a message interleaving basis.

4.5 Packet Switching Techniques in ISDN

ISDN packet switching based on the conventional X.25 protocols in accordance with CCITT Recommendation X.31 (see Sect. 4.4.4) is the standard method adopted for packet mode services in the first phase of ISDN. The main benefit of adapting X.25 to the ISDN stems from the fact that it has reached a high degree of maturity and that the investment in X.25 customer and network equipment can continue to be used with ISDN.

As X.25 had originally been designed for dedicated packet networks using inslot signaling, it is not suitable for true integration into the ISDN. It seems desirable, however, that the consistent separation between signaling (control plane) and the transfer of user information (user plane) which is practised for circuit-mode services in the ISDN should also be extended to packet-mode services in the ISDN. A major advantage is the possibility of establishing common call control for all types of bearer services, including uniform handling of supplementary services.

In the following, two new packet switching techniques will be described that have been specifically designed for ISDN [4.49, 4.50]:

- The Frame Mode Bearer Services (FMBS) and
- the Asynchronous Transfer Mode (ATM).

A comparison of the respective protocol architectures of FMBS (Fig. 4.33a) on the one hand and ATM (Fig. 4.33b) on the other with that of conventional X.25 packet switching (Fig. 4.30) reveals their main differences:

- Both Frame Mode Bearer Services (FMBS) and Asynchronous Transfer Mode (ATM) use a modified version of the D-channel protocol for setting up and clearing down virtual connections, i.e. use of *outslot signaling with single-step call establishment based on Q.931 mod(ified)* instead of the X.25 packet level call control procedures. Apart from the common principle of outslot signaling, FMBS and ATM follow very different approaches for multiplexing and switching of virtual connections (see also Fig. 4.34):
- In Frame Mode Bearer Services (FMBS) *multiplexing and switching of virtual connections (i.e. frame mode connections) takes place at Layer 2,* with a symmetrical variant of HDLC LAP D, i.e. LAP F defined in CCITT Recommendation Q.922 [4.51], being used even in the B and H channels instead of X.25 LAP B in the case of X.25. FMBS perform frame multiplexing according to the multiple link principle by means of a *Data Link Connection Identifier (DLCI)* in the LAP F address field (instead of TEI and SAPI in the case of LAP D); the role of the DLCI corresponds to the logical channel number in X.25. In this way several virtual connections can be identified on a given bearer channel (i.e. D, B, H_0, or H_1).
- Unlike other packet switching techniques, Asynchronous Transfer Mode (ATM) defined in CCITT Recommendations I.150 and I.361 [4.52, 4.53] uses *short fixed length blocks (called cells)* consisting of a 48 octet information field and a 5 octet header (Fig. 4.35). In contrast to the well known Synchronous Transfer Mode (STM), allocation of ATM time slots (i.e. cells) is not performed at regular intervals according to the slot position in an external frame, but in an asynchronous "on demand" manner by means of a *Virtual Channel Identifier (VCI)* contained in the cell header. In order to adapt the flow of meaningful cells (used slots) to the available payload capacity of the transmission system, idle cells (unused slots) must be inserted at the transmit end and marked for extraction at the receive end.

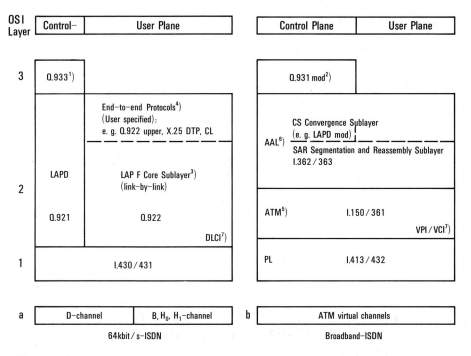

Fig. 4.33a, b. New Packet Switching Techniques in the ISDN.
a Frame Mode Bearer Services (FMBS); **b** Asynchronous Transfer Mode (ATM)

AAL   ATM adaptation layer
CL   Connectionless protocol
DLCl   Data link connection identifier
DTP   Data transfer protocol
LAP F   Symmetrical variant of the HDLC LAP D protocol (modified address field containing DLCl)
PL   Physical layer
VCI   Virtual channel identifier
VPl   Virtual path identifier

[1] modified Q.931 protocol for controlling data link connections identified by a DLCl in the HDLC LAP F address field

[2] modified Q.931 protocol for controlling ATM virtual channel connections (VCC) and ATM virtual path connections (VPC, i. e. globally handled bundles of virtual channels).

[3] service-independent frame relaying mechanism performed in every FMBS exchange.

[4] the common core functions can be supplemented with service-specific and user-selectable end-to-end procedures, e.g. for error recovery (core-and-edge concept).

[5] service-independent hardware-controlled ATM cell transport mechanism executed in every ATM exchange. It includes error control of the cell header but not of the information field.

[6] end-to-end enhancement (core-and-edge concept) of the universal service provided by the ATM Layer in a service-specific way

[7] identifies virtual connections

Technique	Packet Switching			Circuit Switching
	X.25	FMBS	ATM	
Network	64kbit/s–ISDN, PSPDN	64kbit/s–ISDN	B–ISDN	64kbit/s–ISDN
CCITT Recommendation	X.31 (=I.461)	I.122	I.121	I–Series
Multiplexing and switching (OSI Layer)	Packet (3)	HDLC frame (2)	Cells (1–2)	Time slots (1)
Block length (octets)	variable (≤ 128)[1]	variable (≤ 260)[2]	fixed (48+5)	fixed (8)
Set–up and clearing of virtual connections	inslot X.25 / X.75	outslot Q.933 / ISUP mod	outslot Q.931 mod / ISUP mod	outslot Q.931 / ISUP
Maximum bitrate per connection	64 kbit/s	1.5/2 Mbit/s	150/600 Mbit/s	1.5/2 Mbit/s
suitable for delay sensitive CBR services (e. g. voice, video)	No	No	Yes	Yes
Error recovery	Link–by–link	End–to–End	End–to–End	End–to–End

Fig. 4.34.   Overview of Existing and New Switching and Multiplexing Techniques in ISDN

FMBS   Frame Mode Bearer Service
ATM   Asynchronous Transfer Mode
[1]   Value offered by <u>all</u> networks. Other packet lengths (negotiable between user and network)
        are 16, 32, 64, 256, 512, 1024, 2048 and 4096 octets.
[2]   Default value. Other information field lengths can be negotiated between user and network (e. g. 1598 for LAN interconnection)

4.5.1 Frame Mode Bearer Services

The basic concept of Frame Mode Bearer Services (FMBS) is described in CCITT Recommendation I.122 [4.54]. While *Q.931 type D-channel signaling procedures* are used in the ISDN exchanges to establish frame mode connections, *frame handlers* have to be added as specific switching equipment for switching frame mode connections.

Regarding the *control plane*, the existing Q.931 basic call control procedures for circuit-switched calls can be retained for frame mode calls, if information specific to FMBS is included in the information elements (Fig. 4.33a). These modifications concern e.g. the use of a new *Data Link Connection Identifier information element* in the SETUP message, the introduction of a new Transfer Mode (= Frame Mode) in the *Bearer Capability information element*, etc. The DSSI signaling specification for FMBS will be published as CCITT Recommendation Q.933 [4.55].

As for the *user plane*, the network does not fully exploit the HDLC (LAP F) protocol, in contrast to X.25. Instead, the LAP F functionality which is terminated in every FMBS exchange may be restricted to a basic subset – see LAP F *Core Sublayer* in Fig. 4.33a. These core functions provide a low-cost application-independent frame relaying mechanism on a link-by-link basis, comprising

- frame delimiting, alignment and transparency;
- frame multiplexing/demultiplexing using the DLCI in the frame address field;

- detection of transmission errors;
- congestion control functions (see CCITT Recommendation I.370 [4.56]).

By supplementing the common core transport functions with *user specified and service-dependent end-to-end protocols* (e.g. Q.922, X.25 data transfer protocol, connectionless protocols) above the frame relaying mechanism, it will be possible to support different applications: *core-and-edge concept.*

The network may also terminate additional functions above the LAP F core sublayer. Depending on the degree of protocol termination in the network, the following types of FMBS Bearer Services have been defined:

- With the *Frame Relaying Bearer Service* only the LAP F core functions are terminated in the network (see CCITT Recommendation I.233.1 [4.57].)
- The *Frame Switching Bearer Service* provides *full* LAP F functions in the network (see CCITT Recommendation I.233.2 [4.58]).

4.5.2 Asynchronous Transfer Mode

According to CCITT Recommendation I.121 [4.59] summarizing the basic principles for the evolution of the 64 kbit/s ISDN into *Broadband ISDN* (B-ISDN; see Sect. 6.1), Asynchronous Transfer Mode (ATM) will form the basis of B-ISDN. In the meantime, the CCITT has adopted a set of twelve additional Recommendations which contain detailed descriptions of B-ISDN in terms of the ATM Layer, the ATM Adaptation Layer (AAL), the user-network interface as well as service, network and OAM aspects [4.60].

In contrast to FMBS, ATM can also be applied to constant bit rate services with more stringent delay requirements due to using *short fixed-length cells* (Fig. 4.35). Similarly to FMBS, ATM also follows the *core-and-edge approach* by distributing the overall ATM functionality into the ATM Layer and the ATM Adaptation Layer AAL (Fig. 4.33b):

- The *ATM Layer (ATM)* contains the service-independent ATM core functions associated with the cell header (see CCITT Recommendations I.150 and I.361 [4.52, 4.53]). It provides a link-by-link cell transfer capability resulting from processing of the cell header and cell routing in every ATM switch.
- The *ATM Adaptation Layer (AAL)* enhances the universal service provided by the underlying ATM Layer in an application-dependent way (see CCITT Recommendations I.362 and 363 [4.61, 4.62]). The AAL protocols are implemented end-to-end between the endpoints of the ATM connection, e.g. between the ATM terminal equipment or terminal adaptors and STM/ATM interworking units in the network. Thus, the ATM exchanges are not burdened with service-specific functions.

ATM header functions shown in Fig. 4.35 are e.g.:

- *Header error control (HEC)* is used to minimize misdelivery and loss of cells in the ATM layer.

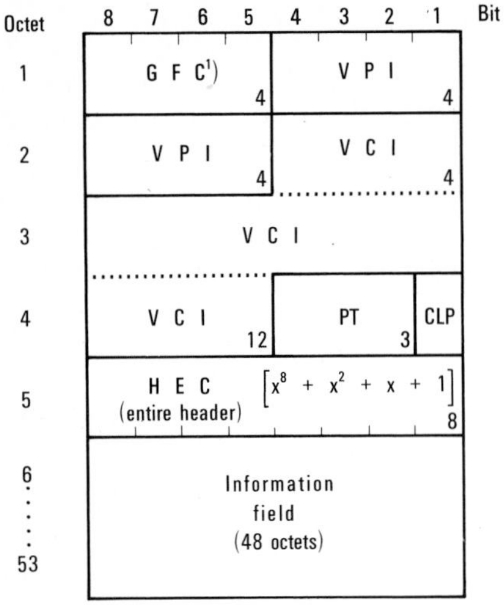

Fig. 4.35. Octets 1–5: ATM cell header.

CLP	Cell Loss Priority	
GFC	Generic Flow Control	
HEC	Header Error Control [CRC-Polynom]	
PT	Payload Type	

VCI	Virtual Channel Identifier
VPI	Virtual Path Identifier
[1]	at UNI only (NNI: 12 bit VPI)

- A *Virtual channel identifier* (*VCI*) refers to a single virtual connection that is switched and routed individually. By means of the *virtual path identifier* (*VPI*), on the other hand, a bundle of virtual connections can be globally identified and switched (virtual trunk).
- The *payload type* (*PT*) indication makes it possible to distinguish within a virtual connection between user information and network information (e.g. OAM cells).
- By means of the *cell loss priority* (*CLP*) the user may indicate to the network on a cell-by-cell basis which cells within a given connection are more sensitive to loss than others; low priority cells will be discarded first under congestion conditions.
- *Generic Flow control* (*GFC*) can be used at the user-network interface to assure proper access of various terminals to a common transmission medium (e.g. bus or ring).

Examples of *ATM Adaptation Layer* (*AAL*) *functions* include

- *error control* of transmission errors affecting the cell information field,
- *segmentation and reassembly* of higher-layer information into or from ATM cell sequences,
- handling of *cell loss and cell misinsertion* in the ATM layer,
- *transfer of timing information* in the case of constant bitrate services (to cope with cell delay jitter introduced by the ATM Layer).

In the *ATM control plane* (Fig. 4.33b), both the signaling application protocol (Q.931) and the signaling data link protocol (Q.921) need to be adapted to the ATM environment. As the total payload capacity of the B-ISDN user-network interface (B-UNI) is subdivided exclusively into ATM virtual channels, virtual channel connections are also used for signaling information transfer. In comparison to the 64 kbit/s ISDN, LAP D (TEI) multiplexing of signaling information to/from several terminals is replaced by multiplexing of ATM *signaling virtual channels (SVCs)*. Typically, one SVC is temporarily allocated to each signaling endpoint in each direction for point-to-point signaling. Accordingly, there is a broadcast SVC which corresponds to the global TEI of LAP D. Management of SVCs across the B-UNI interface is performed via a permanent *meta-signaling channel* for each direction. As a result of partial replacement of the original LAP D functionality by the ATM layer (multiplexing and management of signaling connections) as well as by the SAR sublayer (frame delimiting), only the remainder of the LAP D functionality needs to be implemented in the service-specific CS sublayer for signaling (see LAP D mod in Fig. 4.33b).

4.5.3 Conclusion and Outlook

Both Frame Mode Bearer Services (FMBS) and Asynchronous Transfer Mode (ATM) have been defined with the objective of creating a packet switching technology which is better suited to the ISDN environment than X.25. This has led to conceptual commonalities such as *outslot connection establishment* and the *core-and-edge type protocol architecture* in the user plane (made possible by lower error rates in digital networks). The essential difference is that FMBS uses existing HDLC technology within the framework of the 64 kbit/s ISDN whereas ATM has been defined by CCITT as the future base technology for the evolution towards Broadband ISDN [4.63, 4.64].

Due to the limited bitrate of the underlying STM bearer channels, the principal field of application for FMBS is *data communication up to 1.5/2 Mbit/s*. ATM, on the other hand, makes it possible to exploit the payload capacity of the 155 or even 622 Mbit/s B-ISDN user-network interface, and can therefore be applied to a much wider range of narrowband and broadband applications. As a result of using short fixed-length cells, it can be regarded as combining the relative merits of both circuit switching and conventional packet switching in one *universal technique*, i.e. better time transparency as well as channel bit rates and channel structures on demand. ATM can therefore be applied to *constant bit rate services*, requiring fixed reservation of cells (e.g. voice, circuit emulation), as well as to *variable bit rate services*, employing statistical multiplexing.

From a comparison of the respective user plane protocol architectures of FMBS and ATM in Fig. 4.33, it follows that FMBS data link connections can easily be carried on ATM virtual channel connections without impact on the ATM network. This can be achieved by defining an ATM Adaptation Layer (AAL) protocol type capable of providing the Frame Relaying LAP F core service.

The current set of thirteen CCITT Recommendations on the ATM-based Broadband ISDN which will be supplemented by further specifications (e.g. on signaling) by the end of 1992, can be expected to pave the way for the introduction of ATM technology in the 1993 to 1995 time frame. In view of the bitrate and flexibility limitations of FMBS in comparison with the future-proof ATM technology, potential FMBS applications (e.g. LAN/MAN interconnection) will be likely to migrate to ATM networks as soon as they become available [4.65].

5 ISDN Terminals

5.1 Introductory Remarks

The services of a communication network are accessed via the terminal equipments connected to the user-network interfaces (see Sect. 4.2). The term *terminal equipment* covers terminals for interpersonal communication (e.g. the telephone), terminals for communication between user and data processing systems, and also data processing systems themselves. As the integration of services does not give rise to any special considerations for data processing systems, this chapter therefore deals solely with terminals enabling the human user to gain access to ISDN services.

Terminal equipments are not normally regarded as component parts of a communication network, at least from an administrative or legal standpoint. The network usually ends at the user-network interface, as the term *network termination* (see Sect. 4.1.1) implies. Nor are terminals always component parts of a communication service (cf. Sect. 2.1). Nevertheless, the potential for service integration in the ISDN provided by the service-independent, universal user-network interface (see Sect. 4.2.2) will have considerable effects on the evolution of terminals.

The ISDN-generated impetus for terminal innovation is mainly directed at the terminal functions required for exploiting the new ISDN capabilities. However, ISDN terminals will also exhibit other advanced features generally enhancing user convenience.

An important type of terminal made possible by the ISDN is the *multi-service terminal* (often called a "multifunction terminal"). The multi-channel ISDN access using a common directory number for all channels and with automatic, service-specific terminal selection (see Sect. 4.3.3.2) enables more than one information type to be used alternately or simultaneously for communication both in the office and at home. In many cases this type of multiple communication is only practicable with multi-service terminals (cf. Sect. 5.4).

Due to the expanded range of functions of the new ISDN terminals, every effort must be made to ensure that the terminals are easy to use with all their functional complexity.

Even the terminals available today have many different user interfaces for information input, information output and operation.

Telephones, for example, use different numbers of keys assigned in different ways for the input of signaling information as well as different tone generators and displays for output of that information. With the development of new terminals, especially multi-service terminals, this multiplicity of user interfaces may increase still further.

In order to avoid burdening the public network with this increasing diversity of user interfaces, and also to ensure that the development of new user interfaces is not held back by the public network, the signaling events at the ISDN user-network interface are described as functionally as possible, in other words as independently as possible from the concrete physical input/output units of a specific terminal.

According to the *functional protocol concept* (see Sects. 4.3.5.1 to 4.3.5.5), it is the terminal itself which makes the conversions between the events at the man-machine interface and the functional signaling events at the user-network interface (signaling adaption, Fig. 5.1a). This allows each terminal to provide the interface best suited to its user without individual requirements on the network.

In some national networks it is intended to use (at least in an interim phase) a "stimulus protocol" (cf. Sect. 4.3.5.6) in which the man-machine interface is controlled directly by the network. This may be applied in connection with terminals that are used in large numbers (especially telephones). The terminal reports events at the man-machine interface, such as keyboard inputs, directly to the network as "stimuli" without processing in the terminal (Fig. 5.1b). The stimulus mode terminal has greater flexibility because features can be introduced or changed centrally by the network without modification to the terminal. However, for this purpose program-specific and equipment-specific data must be maintained in the network for each terminal type. A CCITT Recommendation [5.1] exists for the functional protocol and the basic concepts of the stimulus protocol. (For stimulus and functional protocols, see also Sect. 4.3.5.6.)

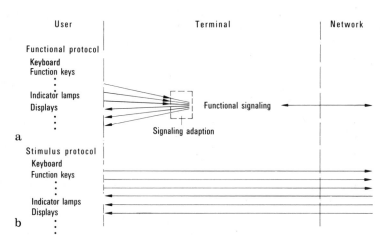

Fig. 5.1a, b. Man-machine Interface at a Terminal and User-Network Interface in the ISDN. **a** With the functional protocol, **b** with the stimulus protocol

5.2 Basic Features of an ISDN Terminal

As described in Sect. 4.2.2, with the basic access the ISDN offers the user an interface with bus capability permitting the operation of several terminals or of multi-service terminals at a *single* user access via parallel-connected sockets (Fig. 5.2).

The functions which the terminals must incorporate to enable them to be operated at this user-network interface are described below. Fig. 5.3 shows the connection unit required in the terminal for the ISDN basic access.

The *line coupler* module provides the electrical connection to the four wires of the interface, e.g. using transformers. It is also responsible for extracting the feeding current supplied across the interface. The *power feeding* module conditions the current from the network termination NT or – if the local AC mains power voltage fails – from the exchange (this implies e.g. voltage stabilization). Where comparatively high power is required (e.g. for operating data terminals with screens), this module enables the terminal to be connected to the AC mains supply.

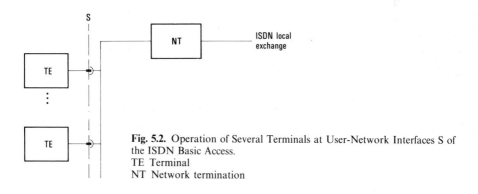

Fig. 5.2. Operation of Several Terminals at User-Network Interfaces S of the ISDN Basic Access.
TE Terminal
NT Network termination

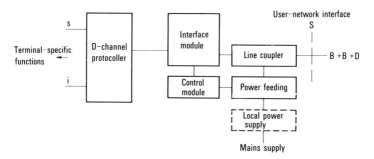

Fig. 5.3. Connection Unit in the Terminal for the ISDN Basic Access.
s Signaling data
i User information

The *control* module contains the functions necessary for controlling the user-network interface e.g. for initializing its operating status and for access to the common signaling channel if the terminal is operated on a passive bus (cf. Sect. 4.3.3.3). The *interface module* is primarily responsible for generating the multiplex structure (B + B + D) at the interface. The *D-channel protocoller* handles signaling with the exchange using the protocol provided for the D-channel, layers 2 and 3 (see Sect. 4.3).

The above summarizes the functions generally required for connecting a terminal to the ISDN. Additional functions are necessary depending on the specific task of each terminal; these are explained in the following sections.

5.3 Single-Service Terminals Connected to the ISDN

This section describes terminals which provide communication only in one information type, e.g. telephony only or text/data communication only.

5.3.1 ISDN Telephone

The single-service terminal which will probably see the largest number of changes due to the influence of the ISDN, in terms of both user functions and appearance, is the telephone [5.2, 5.3]. ISDN telephones can have different levels of sophistication.

Figure 5.4 shows the basic structure of the ISDN telephone. An important feature is the *alphanumeric display* (e.g. one row with 16 positions) which can be

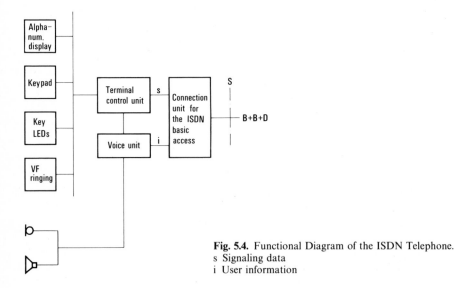

Fig. 5.4. Functional Diagram of the ISDN Telephone.
s Signaling data
i User information

used to indicate directory numbers and other information (cf. Sect. 2.3.3). The *keypad* contains the pushbuttons for connection set-up and function keys for activating supplementary services of the ISDN telephone service.

A basic telephone will only have a small number of function keys for selected supplementary services. The substantially larger number of function keys on advanced telephones can be arranged in blocks according to function groups: one block of name keys, a second containing the keys for specific call establishment functions such as call forwarding/call diversion and completion of calls to busy subscribers. To a third block can be assigned supplementary services providing the subscriber with information, such as registration of incoming calls and advice of charge (cf. Sect. 2.3.3, Table 2.2). A further facility on advanced telephones might be a card reader for identifying the user for different purposes such as assigning charges to individual users of the line, or the activation of the "follow me" function at a visited telephone.

An incoming call is indicated acoustically with *VF ringing.*

The *voice unit* converts the speech signals from analog to digital form and vice versa (PCM encoding/decoding; in some cases using other codes such as ADPCM (Sect. 7.2.1) in order to obtain enhanced transmission quality). The *terminal control* contains the scanning function for the keys and the controller for the telephone's alphanumeric display. B-channel selection also takes place here as initiated by the exchange (see Sect. 4.3.5).

Telephoning without a handset – *handsfree talking* – leaves the speaker's hands free during a call and also allows active participation by other persons. In the analog telephone network, the imperfect attenuation of the hybrid circuit responsible for separating the transmission directions of the speech signals causes the outgoing signal to be fed back via the telephone loudspeaker. This can cause "singing". To prevent that effect, the transmission direction inactive at any one time must be artificially attenuated and, at the same time, in order to obtain an adequate receive volume, the other direction must be amplified depending on the line attenuation. With analog handsfree speaking, the effect of this voice-controlled operation may be so great that the impression of "half-duplex" communication is produced. With the all-digital connections in the ISDN with separate go and return paths, there is no electrical coupling at a hybrid. However, unwanted feedback may be produced by acoustic coupling at the distant end; for this reason voice control is also necessary for digital connections, though in reduced form only, due to the lack of attenuation inherent in digital transmission. Consequently, normal duplex operation is largely maintained.

With digital handsfree talking [5.4], all control functions are performed digitally: processing and attenuation of the digitized voice signal, speech direction recognition, automatic volume control to the level set by the user, and selection of the different operational states: handsfree speaking/open listening/handset.

Figure 5.5 contains a block diagram showing the handsfree equipment with digital voice control. The digital attenuator provides the digital code words of

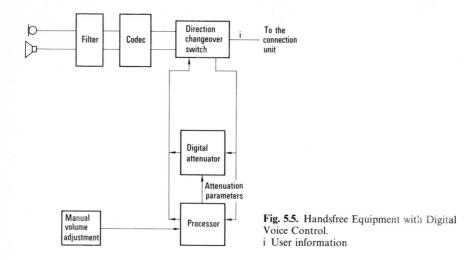

Fig. 5.5. Handsfree Equipment with Digital Voice Control.
i User information

attenuated PCM-encoded speech for both directions. The code words are derived from the original code words in accordance with the intended attenuation. The processor determines the necessary attenuation values by comparing the relative volumes in both transmission directions.

The use of digital control technology has advantages for the manufacture of handsfree talking equipment (compactness by use of integrated circuits) and further improves the quality of handsfree operation through rapid control response and through precise counterbalancing of the transmit and receive channel attenuation.

A number of standards for ISDN telephones exist or are in preparation. For example, the future European standard NET 33 comprehensively describes the technical characteristics which ISDN telephones must possess [5.5].

5.3.2 Terminals for Non-Voice Communication in the ISDN

For the user, the most important new features of terminals for text and data communication as well as facsimile transmission [5.6] in the ISDN is rapid information transfer on the B-channel.

As with the ISDN telephone, the ISDN supports communication activities (call establishment, etc.) for the user of terminals for text and data services. For example, these terminals can also have function keys and an alphanumeric display like the ISDN telephone, in addition to their screen. However, not all the ISDN supplementary services for call establishment, etc. are provided on non-voice terminals, as not all the supplementary services are applicable (cf. Table 2.2).

Figures 5.6 and 5.7 show the basic structure of ISDN terminals for text and facsimile communication.

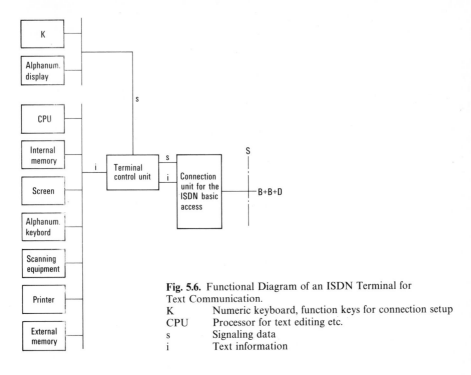

Fig. 5.6. Functional Diagram of an ISDN Terminal for Text Communication.

K   Numeric keyboard, function keys for connection setup
CPU   Processor for text editing etc.
s   Signaling data
i   Text information

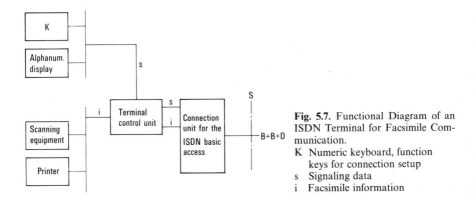

Fig. 5.7. Functional Diagram of an ISDN Terminal for Facsimile Communication.

K Numeric keyboard, function keys for connection setup
s Signaling data
i Facsimile information

A typical ISDN terminal for text and image communication is the "mixed mode" terminal for the ISDN textfax service (see Sect. 2.3.1.2), as it is only the high transmission rate of 64 kbit/s and the correspondingly short transmission time for the facsimile parts of documents that make effective communication with text and image possible.

Admittedly, the cost of the high-resolution screen with large refresh memory required to display an A4 facsimile with sufficient sharpness, together with the

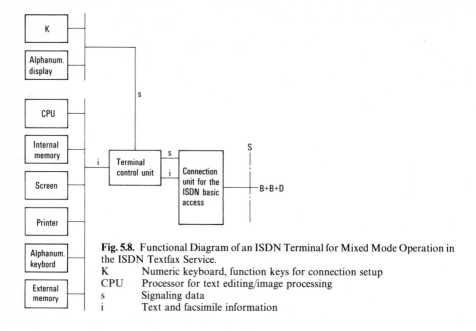

Fig. 5.8. Functional Diagram of an ISDN Terminal for Mixed Mode Operation in the ISDN Textfax Service.

K	Numeric keyboard, function keys for connection setup
CPU	Processor for text editing/image processing
s	Signaling data
i	Text and facsimile information

scanning and printing equipment, is not inconsiderable. Figure 5.8 shows the basic structure of an ISDN mixed mode terminal. At its highest level of sophistication, a terminal of this type can read existing documents in such a way that the textual material is represented in alphanumeric code (for which automatic character recognition is required) and only the pictorial material appears as facsimile, if appropriate in vector graphics. Thus the volume of information to be stored and transmitted can be reduced to a minimum by using the most economical encoding method in each case.

On-screen text editing is carried out using normal text editing software, as the text is available in character representation. The printer is designed to output both text and facsimiles. Like all text editing equipment, the terminal is equipped with an external memory (e.g. floppy disk).

For terminals to be used in connection with ISDN teleservices for text and data communication as well as facsimile transmission (in CCITT: telematic services), internationally standardized communication protocols exist (cf. Sect. 2.1). Whereas the higher, application-oriented layers of these protocols (corresponding to the OSI reference model) can basically be defined independently of the ISDN, the lower, transport-oriented layers of the protocols are naturally ISDN-specific. For these layers of the communication protocol, the CCITT has published a Recommendation T.90 [5.7].

In order to allow unrestricted data communication worldwide, the difference between the digital systems (for an interim period, 56 kbit/s basic channel only usable for non-voice e.g. in the USA, as opposed to 64 kbit/s e.g. in Europe) must be overcome. This may have implications for terminal equipment.

5.4 Multi-Service Terminals

Multi-service terminals are equipments which have communication capability in more than one information type, either alternately or simultaneously [5.3]. In principle this is also possible with several dedicated terminals; however, the integration of several information types in one unit has decisive advantages:

- A multi-service terminal occupies much less space than the corresponding number of single-service terminals.
- Change of service during a call and simultaneous communication in two services are easier (this includes voice annotations to text and image).
- Only one set of call-establishment equipment is required.
- With a multi-service terminal for voice and data, telephone communication can be improved by the screen being used to display not only data but also a locally stored telephone directory.

5.4.1. Multi-Service Terminals for Telephony and Telematic Services

An initial example of a multi-service terminal is an integrated voice and data terminal (IVDT) which can be used for ISDN telephony and videotex alternately, or even simultaneously, thanks to the two B-channels in the ISDN basic access. It is equally suitable for use in the office and in the home. Figure 5.9 shows the basic design of this IVDT. As in all multi-service terminals, the functional units shown in the diagram are physically grouped in accordance with operational and user-oriented criteria. Telephoning is enhanced by useful features such as a personal telephone directory in the internal memory and user prompting [5.8, 5.9], the screen being used for information output in connection with these features. A particularly effective support facility for telephone traffic is automatic initiation of call set-up where the cursor is used to mark the desired directory number in the directory displayed on the screen: dialling is then started

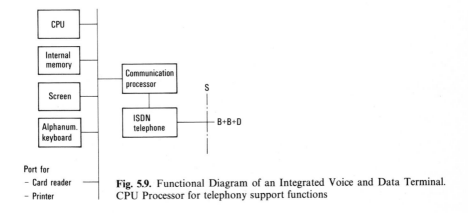

Fig. 5.9. Functional Diagram of an Integrated Voice and Data Terminal. CPU Processor for telephony support functions

by pressing just one additional key. This is a typical example of avoidance of "media discontinuity": manual entry of the number after consulting the directory is no longer necessary.

The IVDT has an alphanumeric keyboard for using the videotex messaging service and for local data entry. Other local functions are conceivable, such as keeping an appointment diary and operating a card reader. The information content of one's own IVDT can be stored on the card and loaded to other IVDTs of the same type so that, for instance, one's personal telephone directory can be used there, too.

In most cases, the IVDT will support a connected printer.

Another advanced multi-service terminal is an IVDT in the form of a (PC-based) office workstation. As well as optional utilization of several telecommunication services, workstations provide a range of local functions similar to that found in personal computers. Figure 5.10 shows the basic layout of a workstation.

In addition to telephony, it is possible to communicate with public and private videotex centers, with public and private message handling systems and with special application programs in data processing systems. Workstations can also operate with teletex protocols and use the teletex service.

Local functions can include

- text processing
- pocket calculator functions
- loading and running application programs
- office service functions such as appointment diary, reminders, work schedule
- memory functions such as storage with automatic retrieval, or storing personal notes
- support of telephony

The ISDN workstation also allows simultaneous voice and non-voice communication (e.g. for an enquiry to a database during a telephone conversation).

Like personal computers, workstations provide connection facilities for various peripherals such as printers and external memories.

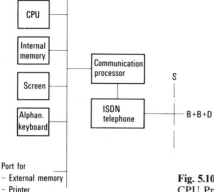

Fig. 5.10. Functional Diagram of an ISDN Workstation. CPU Processor for text and data functions

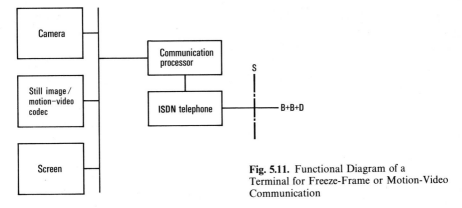

Fig. 5.11. Functional Diagram of a Terminal for Freeze-Frame or Motion-Video Communication

5.4.2 Multi-Service Terminals for Telephony and Image Communication

This Section deals with the subject of terminals from the point of view of future services for still image and motion-video communication in the 64-kbit/s ISDN. Terminals for video communication are here classed as multi-service terminals, because the exchange of still and moving images will generally be accompanied by voice communication.

In terms of their utilization, terminals for video communication are typical ISDN terminals, because video communication, particularly motion-video, is only now becoming feasible due to the high transmission rate in the ISDN.

However, in order to offer acceptable video communication at the ISDN transmission rates of 64 kbit/s or at best 2×64 kbit/s, the amount of data to be transmitted has to be reduced using special coding methods (by eliminating redundant and irrelevant information). For still image communication both the "sequential" and the "progressive picture buildup" method has proved particularly effective. With the technically simpler "sequential buildup" technique a full color TV still image is transmitted and displayed in a few seconds. With the more complex "progressive buildup" technique the image is transmitted and displayed first but almost instantaneously with low spatial resolution, and the fine details are progressively added in the next few seconds by gradually increasing the resolution of the image.

International standards for still image and motion-video communication equipment are in preparation, with particular emphasis placed on the coding methods. As far as the still image is concerned, the combined efforts of the CCITT and ISO have already produced concrete results [5.10]. For motion video (in the 64-kbit/s channel), the CCITT has published recommendations [5.11].

The functional diagram for a still image and motion-video communication terminal is shown in Fig. 5.11. As the ISDN telephone additionally controls the picture components, it is provided with special keys for this purpose.

For further details on video communication in the ISDN, refer to [5.12 to 5.14].

6 Switching in the ISDN

6.1 Introduction

The cost-effective realization in LSI chips of digital time-division multiplexing and speech digitization (64 kbit/s pulse code modulation–PCM–in accordance with CCITT Rec. G.711 [6.1]) has resulted in a new concept for switching equipment.

Typical of the new generation of equipment is that used in modern data networks [6.2, 6.3] and for the digitized telephone network (Integrated Digital Network IDN) [6.4]. The digital telephone network develops from the existing analog network as a consequence of the cost benefits of using digital components. The evolving network is characterized by the integration of digital transmission (cf. Sect. 7) and digital switching. As already discussed in Sects. 1.6, 1.7 and 3.3, the digitized telephone network forms the basis for the ISDN [6.5 through 6.7].

Digital technology not only allows an extended range of user facilities to be provided (see ISDN supplementary services in Sect. 2.3.3 and [6.8]) but also enables all the services to be handled via the same subscriber access. An important principle in ISDN is the separation of service aspects on the one hand and network aspects on the other (Fig. 6.1). The teleservices and bearer services are implemented within the network by means of appropriate *network capabilities* (see CCITT Recs. I.210 [6.9] and I.310 [6.10]) with low layer capabilities (LLCs) in protocol layers 1 to 3 and high layer capabilities (HLCs, relating to teleservices only) in layers 4 to 7. LLCs comprise "basic low layer functions" for setting up and clearing down ISDN connections and "additional low layer functions" for handling supplementary services. Basic low layer functions are determined essentially by the *ISDN connection types* defined in CCITT Rec. I.340 [6.11]. Like the ISDN services, the connection types are described in terms of appropriate attributes (Table 6.1). Dominant connection attributes include the mode (circuit or packet switching) and the bit rate for transferring user information.

The essential functional elements of a public ISDN from the point of view of switching are shown in Fig. 6.2:

- *ISDN local exchanges* with digital subscriber lines (see Fig. 3.1),
- *Common channel signaling capabilities* (CCITT signaling system No.7) for transferring signaling information between exchanges,

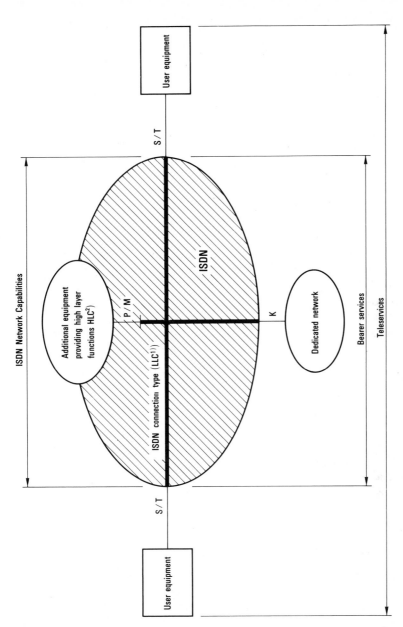

Fig. 6.1. Relationship Between Services and Network Capabilities of an ISDN.
S, T, K, P, M Reference points
[1] Low Layer Capabilities (relating to layers 1-3)
[2] High Layer Capabilities (associated with layers 4-7): specialized resource within the ISDN (P) or specialized service provider outside the ISDN (M)

Table 6.1. Attribute Values for ISDN Connection Types (in accordance with CCITT Rec. I.340)

Attribute	Values[a]
Information transfer mode	– circuit-switched (cs) – packed-switched (ps)
Information transfer rate	– cs: 64 (B), 384 (H0), 1536 (H11), 1920 kbit/s (H12) – ps: packet throughput
Information transfer susceptance	– unrestricted digital information (i.e. any bit sequence) – speech[b] (digital encoding, e.g. A law, μ law) – 3.1 kHz audio[c] (e.g. voice-band data)
Establishment of connection	– switched[d] – semi-permanent[e] – permanent[f]
Symmetry (of information transfer)	– unidirectional – bidirectional symmetric – bidirectional asymmetric[g]
Connection configuration – topology – dynamics	 – point-to-point, multipoint – concurrent, sequential, or/add remove[h]
Structure (of user information) – layer 1 – layers 2 and 3	 – 8 kHz integrity[i], unstructured – Service data unit integrity, unstructured
Performance – error performance – slip performance	 – CCITT Rec. G.821 (see Sect. 7.7.1) – CCITT Rec. G.822 (see Sect. 7.7.2)

[a] applying to the overall ISDN connection consisting of access and transit connection elements
[b] bit manipulation may be applied (see Sect. 7.2.1)
[c] suitable for modem signals
[d] on demand set-up in response to signaling information from the subscriber
[e] connection passes through a switching network (exchange)
[f] connection uses the transmission network only (by-passing of exchanges)
[g] i.e. different information transfer rates in the two directions
[h] refers to the time sequence in which connection elements are set up and released
[i] bits submitted within a demarcated 125 μs interval are delivered within such an interval

- physical connections with *circuit switching*,
- virtual connections with *packet switching*,
- *specialized equipment* for additional functions.

A large number of ISDN supplementary services can be provided with the aid of stored program control (SPC) exchanges in conjunction with the principle of outslot signaling, i.e. separate channels for transmitting user information and signaling information (see Sect. 2.3.3). Control of these ISDN features such as completion of calls to busy subscribers (CCBS) resides almost exclusively in the *ISDN local exchanges* (see Fig. 3.1). In the transition from the digital telephone network to the ISDN, therefore, it is the local exchanges which primarily have to be upgraded. This involves not only modifying the hardware to provide digital transmission on subscriber lines (see Sect. 4) but also expanding the (switching) software (see Sect. 6.2).

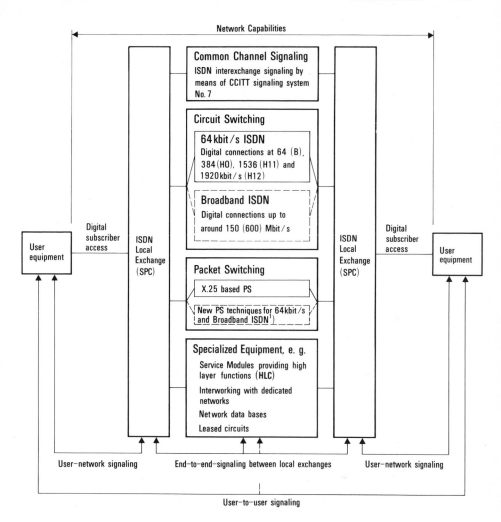

Fig. 6.2. Basic Architectural Model of an ISDN.
LE   Local exchange
SPC   Stored Program Control
HLC   High Layer Capabilities (associated with layers 4-7)
[1] see Sect. 4.5

A great many ISDN-specific functions, processes and supplementary services have to be handled in the ISDN local exchanges (see Sect. 2.3.3), whereas the transit exchanges are affected by ISDN to a much lesser degree. Consequently, the *common channel signaling* system has to provide the additional facility of signaling between the local ISDN exchanges without the need for the signaling information to be processed in the transit exchanges involved in the connection. In other words, end-to-end signaling between originating and terminating ISDN exchanges is to be provided (see Sect. 6.3). In addition, within CCITT signaling system No. 7 [6.12 to 6.15], the traditional separate signaling procedures for speech (see Telephone User Part TUP [6.16 to 6.20]) and text/data (see Data User Part DUP [6.21]) have to be replaced by a *single* universal integrated services signaling protocol (see ISDN User Part in Sect. 6.3 and [6.22 to 6.27]).

The starting point for ISDN is a basic system with circuit switching based on the digital telephone network and featuring the following principal functions (Figs. 6.2 and 6.3):

● switching of digital B channel connections (connection control), and
● signaling functions (call control).

With the transition from the digital telephone network to ISDN, the circuit-switched digital connections of the trunk network, originally provided for transmitting PCM-encoded telephone signals, will become universal transparent 64 kbit/s user-to-user connections which in turn will form the basis for a wide range of text and data services (see Unrestricted Digital Information in

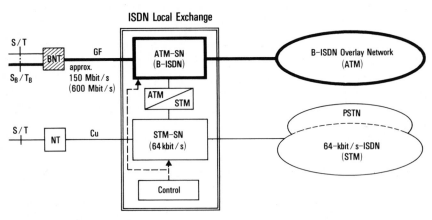

Fig. 6.3. Evolution of ISDN Towards Broadband ISDN.

———	64 kbit/s ISDN
▬▬▬	Broadband ISDN (B-ISDN)
- - - -	Signaling (call control)
NT	Network termination
BNT	Broadband network termination
Cu	2-wire copper access line
GF	Glass fiber
PSTN	Public switched telephone network
SN	ISDN switching network
ATM	Asynchronous Transfer Mode (see Sect. 4.5)
STM	Synchronous Transfer Mode

Table 6.1). As an interim solution in some countries while ISDN is being introduced, some ISDN connections may be routed via analog network sections. Moreover, ISDN sections may be routed via digital sections which operate with transcoding to 32 kbit/s ADPCM (see Sects. 7.2.1). Both measures are only applicable if the relevant service permits the analog transmission or the bit manipulation that they involve; this covers speech and voice-band data as shown in Table 6.1 (3.1 kHz audio, and suitable also for modem signals).

Voice and circuit-switched non-voice services require the same basic ISDN functions for signaling (call control), routing etc. available in all ISDN local and transit exchanges. Accordingly common signaling protocols are employed on the user side (see Sect. 4.3) and on the interexchange side (see Sect. 6.3) for ISDN telephony and for circuit-switched text and data services; this is the principle of *single-step call establishment* (see Sects. 4.4.1 and 4.4.2).

The principle of full integration described above applies also to the expansion of the switching equipment of the 64 kbit/s ISDN for use in the *broadband ISDN* at a later stage (Figs. 6.2 and 6.3). This affects the following aspects of the ISDN exchanges (see Sect. 6.2 and [6.28 to 6.30]):

- the *subscriber access*, which must be equipped for optical fibers (see Sect. 7.3.1) instead of two-wire copper access lines, and
- the *switching network*, which must be supplemented with broadband modules (approx. 150 Mbit/s and 600 Mbit/s).

Although full integration of *packet switching* techniques in the ISDN is possible in principle, this must wait for a later phase owing to the considerable effects it would have on the ISDN exchanges (see below). If *conventional X.25 terminals* are connected to the ISDN, the rest of the ISDN, i.e. the basic system, merely functions as a transparent access from the local ISDN exchanges via 64 kbit/s trunk connections to special packet switching equipment which may reside either *inside* the ISDN (see the use of the *ISDN virtual circuit bearer service* described in Sect. 4.4.1) or *outside* the ISDN (see the *access to data transmission services provided by separate packet networks*, discussed in Sect. 4.4.1). In both cases, ISDN user-network and interexchange signaling serve merely to establish the access connection so that for packet-switched bearer services additional service-specific protocols have to be handled "in slot", i.e. in the B channel; this is the principle of *two-step call establishment* with separate call control for circuit and packet switching (see Sects. 4.4.1 and 4.4.4).

The key concept of *new ISDN packet switching techniques* (see Sect. 4.5), which are not based on CCITT Recs. X.25/X.75, involves common call control and uniform signaling protocols for circuit *and* packet switching in accordance with the principles of single-step call establishment and outslot signaling – in other words, the use of modified D channel protocol according to CCITT Rec. I.451 (see Sect. 4.3.5) instead of X.25, and ISDN interexchange signaling ISUP (see Sect. 6.3) instead of signaling based on CCITT Rec. X.75. Two of the advantages of this farther reaching integration of packet switching in ISDN are as follows:

- uniform handling of supplementary services and of operation and maintenance functions for circuit and packet switching;
- possibility of simultaneous use of circuit and packet-switched bearer services in the course of a *single* call (multiservice mode).

For high-layer functions (high layer capabilities) associated with protocol layers 4 to 7 and for certain functions in layers 1, 2 and 3 (low layer capabilities) *specialized equipment* is also required in the ISDN (Fig. 6.2). An example of such a low layer function is *interworking with dedicated text and data networks* (see Sect. 4.4 and Figs. 6.4, 6.7). For reasons of economy and practical implementation, interworking equipment will not be provided in every ISDN exchange but will rather be centralized and assigned to certain exchanges at the higher network levels.

Another example of specialized equipment is a digital cross connect DCC providing for the control of transmission links, e.g. setting up of *leased circuits*. While ISDN exchanges are optimized for establishing circuit-switched connections lasting typically a few minutes, DCC connections are maintained for longer periods of time (from days to years).

Service modules for high layer functions can be implemented as feature nodes FN or as vendor feature nodes VFN, depending on whether value-added services VAS are offered to ISDN subscribers by the network carrier or by independent enhanced service providers (see Figs. 6.5, 6.8). VAS can be characterized as involving storage and processing of user information. Fundamentally, they comprise four main categories: data base access, store-and-forward communication of the message handling type (CCITT X.400 series of Recs.), support services such as electronic online directories (CCITT X.500 series of Recs.), and compatibility services (see Sect. 2.3.1.3 and [6.31]).

The feature nodes (FN, VFN) mentioned above add intelligent functionality to the ISDN; this functionality is offered directly to the user, i.e. enhanced services are offered to the subscriber via the basic services. *Network data bases* (see service control point SCP in Figs. 6.5, 6.8), which can be accessed on a real

---

Fig. 6.4. Functional Structure and Communication Links of an ISDN Local Exchange.

B	B-channel	PH	Packet Handler (see Sect. 4.4)
CAS	Channel associated signaling	PABX	Private Automatic Branch
CCS	Common Channel Signaling		Exchange
CS	Circuit switched user information	PS	Packet switched user information
CSPDN	Circuit Switched Public Data	PSPDN	Packet Switched Public Data
	Network		Network
D	D-channel	s	Signaling information
d	Data (non-voice) information	SCP	Service Control Point
DCC	Digital Cross Connect		(see Sects. 6.1, 6.3.7)
ISUP	ISDN User Part (CCITT	STP	Signaling Transfer Point
	Signaling System No. 7)		(see Sect. 6.3)
IWU	Interworking Unit: Signaling	v	digitized voice information
	conversion from ISUP to X.71	VFN	Vendor Feature Node
LAN	Local Area Network	1 to 5	Communication links via the
p	Packet data (on D-channel)		switching network (see Sect. 6.2.3)

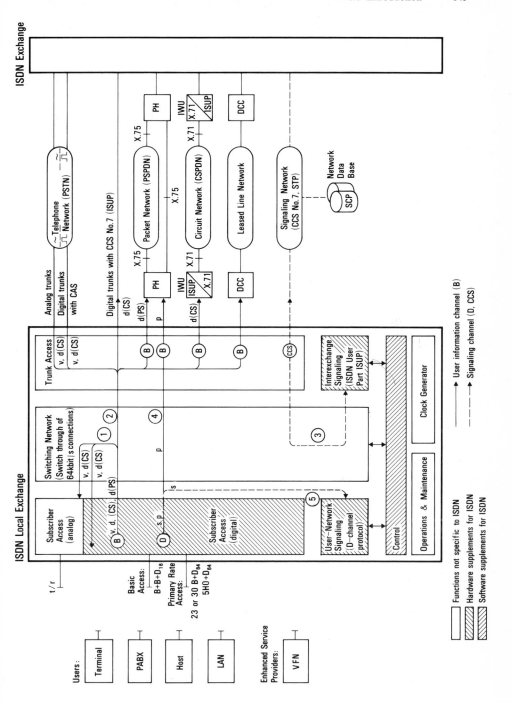

time basis during call control by ISDN exchanges via the common channel signaling network, result in a different kind of internal network intelligence intended for improving the provision of basic services. The main objective of this concept of the *intelligent network* [6.29, 6.32–6.34] is to give administrations a greater amount of flexibility in terms of rapid introduction of new services, even on a trial basis (see Sect. 6.3.7). Since the control logic for handling special calls resides in centralized network data bases, it can be changed without affecting the call processing software in the various ISDN exchanges. Another benefit of the SCP concept is the ability to customize services for unique customer needs including the concept of embedding private networks as logical subnetworks into the resources of the public network; such embedded networks are sometime termed "virtual corporate networks".

6.2 New Demands Placed on Switching Due to Service Integration in the ISDN

The new demands made on switching are mainly attributable to integration of services. To meet these demands, new ISDN-specific functions, evolving from the functional concept of switching in the digitized telephone network, have had to be introduced. These have resulted in corresponding changes or expansions to the hardware and software. Figure 6.4 shows the functional structure of an ISDN local exchange and its surrounding:

- Subscriber access
- Switching network
- Trunk access
- User-network signaling
- Interexchange signaling
- Control
- Operations, administration and maintenance
- Timing and network synchronization.

The ISDN-specific hardware and software functional units are shown hatched. A functional exchange model can also be found in CCITT Rec. Q.521 [6.35].

6.2.1 Subscriber Access

The functional unit for the subscriber access comprises all the layer 1 interface functions, particularly those for connection of digital ISDN subscriber lines. This includes functions for interoperation with transmission equipment (cf. Sect. 7.4.3) as well as functions associated with the interface structures for the *basic access* $(B + B + D_{16})$ and for the *primary rate access* (primarily $23 B + D_{64}$ or $30 B + D_{64}$) already presented in Sect. 4.2, Table 4.3. Figure 6.5 relates to CCITT Recs. Q.5 11 [6.36] and Q.512 [6.37] and contains a summary of the

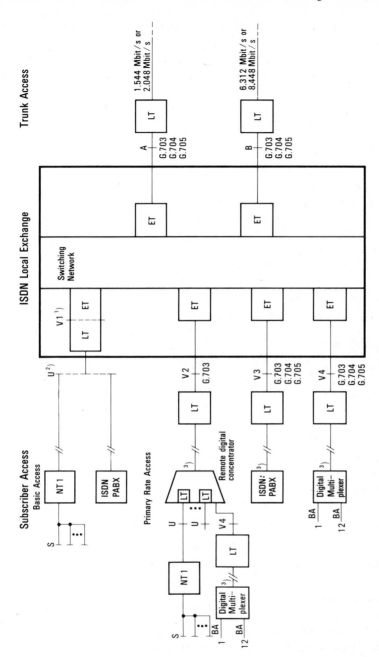

Fig. 6.5. Digital Interface Associated with an ISDN Local Exchange.

BA   Basic Access
ET   Exchange Termination
LT   Line Termination
[1] V1 may be an internal interface in the switching system (system dependent, not subject to standardization by CCITT)
[2] not yet subject to standardization by CCITT
[3] LT not shown

digital interfaces of an ISDN local exchange for both the user and the inter-exchange side (cf. Sect. 6.2.2).

Depending on their size, ISDN private automatic branch exchanges (PABXs) can obtain access to the local exchange either via the basic access (see U interface in Fig. 6.5) or the primary rate access (V3).

Interfaces V1 to V4 separate the transmission functions of the ISDN user-network interface from the switching functions. The physical *line termination* (LT) performs transmission functions whereas the *exchange termination* (ET) logically terminates the subscriber access. The V interface is a functional boundary between ET and LT; it does not necessarily exist as a physical interface, since particularly cost-effective solutions can be provided by an integrated LT/ET implementation. This applies to the ISDN basic access for which transmission and frame structure on the subscriber line are not yet defined internationally but (in some countries) nationally (cf. Sect. 7.4.3 and [6.38]). The transmission functions of the LT in the case of the basic access correspond to those of the NT1 (cf. Sect. 4.1).

Subscriber access functions can also be arranged in remote equipment physically separate from the exchange and connected to it by primary rate time-division multiplex systems: such devices include digital concentrators (interface V2 in Fig. 6.5) and digital multiplexers (interface V4, Sect. 7.4.3); this latter option is of importance in areas of low subscriber density and particularly during the introductory phase of ISDN, in which not all local exchanges will yet have been converted to ISDN (cf. Sect. 3.8).

Interfaces V2 to V4 are based on the same digital transmission techniques (cf. Chap. 7) as that used on the interexchange side. The electrical characteristics of these interfaces, i.e. bit rate, pulse shape, etc., are defined in CCITT Rec. G.703. The functional characteristics of V2 through V4, i.e. particularly the frame structure, are described in CCITT Rec. G.704. CCITT Rec. G.705 contains additional definitions for digital connections terminating on digital exchanges.

In addition to the digital interfaces shown in Fig. 6.5, an ISDN local exchange must also have connection facilities for analog subscriber lines (cf. Sect. 2.3.4), since when a digital ISDN exchange replaces a conventional exchange in the analog network it must take over the analog subscriber lines which remain (Fig. 6.4: analog tip/ring interface t/r). Examples of analog subscriber access functions are: Signaling facilities such as tone generators and digit receivers, conversion between analog and digital voice signals (codec) and conversion between two-wire and four-wire operation.

The functional units for digital, i.e. ISDN subscriber access lines and for analog telephone lines are linked to a switching network by time-division multiplexed (TDM) internal paths as are the corresponding interface units on the interexchange side (see Sect. 6.2.2). The function of the switching network (cf. Sect. 6.2.3) is to through-connect the 64 kbit/s B-channels. For this purpose, a time slot of the incoming internal time-division multiplex system, corresponding to an incoming B-channel, is through-connected to the "set" time slot of an outgoing time-division multiplex system. This requires the functional units for user and interexchange access to adapt the different subscriber lines and interexchange trunks of a combined digital/analog local exchange to the internal digital switching paths.

In addition to digitized voice signals v, the 64 kbit/s user information channels (B-channels) can also transmit circuit switched and packet switched non-voice information (d(CS) or d(PS)) at 64 kbit/s. The D_{16} channel of the basic access can also be used to convey packet data p between ISDN users and the ISDN exchange in addition to signaling information s. The s and p-type information must be separated in the exchange (Fig. 6.4). Whereas the signaling information is intended for the user-network signaling functional unit (cf. Sect. 6.2.4), the p information is forwarded to a Packet Handler PH in a "transparent" manner, i.e. without any processing in the ISDN exchanges through which it passes – e.g. via semi-permanent 64 kbit/s connections used for transferring packet data from a number of subscribers on a message interleaving basis (cf. Sect. 4.4.4).

6.2.2 Trunk Access

The functional unit for the trunk access (Fig. 6.4) comprises all the functions for interconnection towards other exchanges at first and second levels of the digital transmission hierarchy (interfaces A and B in Fig. 6.5). Electrical and functional characteristics of the digital interexchange interfaces A and B are defined in CCITT Rec. Q. 511 [6.36] with references to Recs. G.703, G.704 and G.705 (cf. Chap. 7).

For a comparatively long transition period while analog multiplex systems are still being used at the interexchange level of the telephone network, it must also be possible, for reasons of compatibility with the telephone network, to connect analog transmission systems to an ISDN exchange (Fig. 6.4). For such circuits, a digital to analog conversion facility must be provided; in practice this is implemented by means of a signaling converter combined with a PCM multiplexer (cf. SC-Mux in Fig. 6.7).

Like the corresponding unit for the subscriber access, the trunk access unit is connected to the switching network via time-division multiplexed (TDM) digital paths (Fig. 6.4). In the switching network, the individual 64 kbit/s channels are through-connected. The common 64 kbit/s signaling channels (CCS) used to transfer the interexchange signaling within CCITT Signaling System No.7 are also routed to the switching network in the same way. The common signaling channels are subsequently routed to the functional unit for interexchange signaling via dedicated switching network paths (semipermanent through-connection).

6.2.3 Switching Network

In the digitized telephone network with integrated switching and transmission, the 64 kbit/s channels transmitted on a time-division multiplex basis are switched direct from the TDM systems that are connected to the exchange, i.e. without previous splitting up of the TDM signal into individual 64 kbit/s channels. For PCM voice transmission each channel octet contains a PCM codeword (cf. Sect. 7.2.1).

Thus, in digital time-division multiplex switching, the switching function consists of connecting octets (cf. Sect. 7.2.3, Fig. 7.3) through the switching network from an incoming TDM signal to the relevant outgoing TDM signal (Fig. 6.6). This involves two basically different through-connection methods:

- *Time stages* through-connect an octet by changing the channel time slot between an input TDM system and the associated output TDM system. (Time slot change Ti in Fig. 6.7: An octet received in time slot 10 of incoming system I_n is transferred to time slot 11 of the corresponding TDM system O_n).
- *Space stages* on the other hand through-connect the octets from an incoming TDM system to another TDM system corresponding to the switching destination while retaining their slot. (Change of position in space, Sp in Fig. 6.7: Octet 10 of input system I_1 retains the same phase when transferred to output system O_n).

The switching process generally requires changing the time slot and the position in space of the octet being through-connected. In practice therefore, time stages and space stages have to be combined, e.g. in the form of "time-space-time" switching network arrangements [6.39].

Depending on the type of information (cf. Sects. 6.2.1 and 6.2.2) which is transferred on the 64 kbit/s channels routed via the switching network, the following "exchange connections", as considered in CCITT Rec. Q.522 [6.40] can occur in an ISDN exchange (cf. Fig. 6.5):

(1) B-channel (B) – B-channel (B)
(2) B-channel (B) – interexchange user information channel (B)
(3) Common signaling channel (CCS) – functional unit for interexchange signaling
(4) Signaling channel (D_{16}) – semipermanent feeder channel (B) to a Packet Handler (PH) in the ISDN (see ISDN virtual circuit bearer service solution in Sect. 4.4.4)
(5) Signaling channel (D_{16} or D_{64}) – functional unit for user-network signaling.

For higher-rate channels (cf. Table 4.1) there is also the option of switching more than one 64 kbit/s channel through the switching network in the course of one connection, e.g. five 64 kbit/s channels in the case of the H0-channel with a total bit rate of 384 kbit/s (cf. Sect. 4.2). The H0-channel occupies five 64 kbit/s time slots per 2.048 Mbit/s pulse frame and all of these have to be through-connected to the same address. This requires special arrangements in the switching network to retain the original octet sequence (octet sequence integrity).

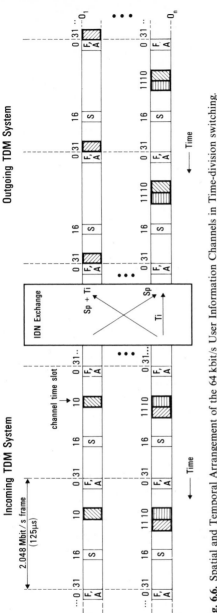

Fig. 6.6. Spatial and Temporal Arrangement of the 64 kbit/s User Information Channels in Time-division switching.

A   Alarm indication and other messages (channel time slot 0)
F   Frame alignment signal (channel time slot 0): zero point for channel numbering
IDN   Integrated digital network: transmission *and* switching of time-division multiplexed digital channels
$I_1 \ldots I_n$ Digital TDM systems – incoming direction
$O_1 \ldots O_n$ Digital TDM systems – outgoing direction
S   Interexchange signaling information (channel time slot 16)
Sp   Change of position in space (space slot change)
Sp + Ti Combined change of space *and* time slot
TDM   Time-division multiplex
Ti   Time slot change

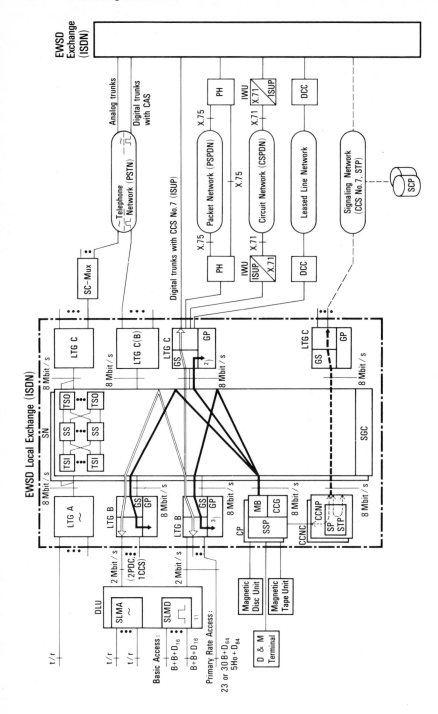

6.2.4 User-Network Signaling

The user-network signaling functional unit – depending on the type of subscriber access – handles ISDN user-network signaling (cf. Sect. 4.3 and Table 6.2) or dialing procedures for analog telephone lines (t/r). This includes receiving and transmitting signaling information as well as conversion of external signaling events into internal messages for the control functional unit and vice versa (Fig. 6.4). The "control" unit assumes the tasks associated with the control of a connection from the incoming side to the outgoing side (cf. Sect. 6.2.5). For example, it determines the path via the switching network between the subscriber access line and the trunk line by evaluating the dial information and sets this path up; in addition, it is responsible for passing on dial information to the interexchange signaling functional unit (cf. Sect. 6.2.6).

Fig. 6.7. System Structure and Communication Links of an ISDN Local Exchange (Siemens system EWSD).

═══	64 kbit/s user information channel (switched)
───	64 kbit/s control channel (semipermanent)
- - - -	64 kbit/s signaling channel CCS (semipermanent)
CAS	Channel associated signaling
CCNC	Common Channel Network Control
CCNP	Common Channel Network Processor
CCS	Common Channel Signaling
CP	Coordination Processor
CSPDN	Circuit Switched Public Data Network
DCC	Digital Cross Connect
DLU	Digital Line Unit
GP	Group Processor
GS	Group Switch
ISUP	ISDN User Part: for interexchange signaling within CCITT Signaling System No.7 (see Sect. 6.3)
IWU	Interworking Unit (see Sect. 4.4.1)
LTG	Line/Trunk Group
MB	Message Buffer
PH	Packet Handler (see Sect. 4.4.4)
PSPDN	Packet Switched Public Data Network
SC-Mux	Signaling Converter – Multiplexer (includes digital to analog conversion)
SCCP	Signaling Connection Control Part (see Sect. 6.3.4)
SCP	Service Control Point (see Sects. 6.1, 6.3.7)
SGC	Switch Group Control
SLMA	Subscriber Line Module Analog
SLMD	Subscriber Line Module Digital
SN	Switching Network
SP	Signaling Point (see Sect. 6.3.1)
SS	Space Stage
SSP	Siemens Switching Processor
STP	Signaling Transfer Point (see Sect. 6.3.1)
TSI	Time Stage Incoming
TSO	Time Stage Outgoing
t/r	analog tip/ring interface

Path of No.7 messages through the CCNC
[1] connection of a digital multiplexer (cf. Fig. 6.6) not shown,
[2] ISUP for link-by-link Signaling (see Sect. 6.3.3),
[3] ISUP for End-to-end Signaling (see Sect. 6.3.3) and SCCP (see Sect. 6.3.4)

Table 6.2. Switching Functions of an ISDN Local Exchange

(1) *Signaling*

- Transmitting/receiving signaling messages in accordance with D-channel protocol (user side) or ISDN User Part (interexchange side),
- Conversion of external signaling events into internal messages in the exchange for the control functional unit and vice versa.

(2) *Access Control*

- Basic call processing (call establishment and clearing),
- Control of call-related ISDN supplementary services (e.g. completion of calls to busy subscribers, in-call modification, etc.),
- B-channel administration,
- Handling of signaling transactions,
- Conversion of functional signaling into stimulus signaling,
- Subscriber messages for registration and cancellation of user facilities,
- Preprocessing of dial information,
- Authorization check for access to ISDN services,
- Recording of charging and traffic data,
- Overload protection,
- Timing functions.

(3) *Switching*

- Analysis of dial information,
- Group selection and zoning,
- Routing,
- Path search in the switching network from an incoming to an outgoing exchange port,
- Through-connection or release of the switching network paths in the course of setting up or clearing down 64 kbit/s connections by means of switching network commands,
- Calculating Charges.

(4) *Interworking with Operations* (e.g. charges, traffic measurement) *and Maintenance* (e.g. fault analysis, recovery)

As already discussed in Sect. 6.2.1, users can also obtain access via remote concentrators (called Digital Line Units DLU in the EWSD system) connected to the local exchange via primary rate time-division multiplex systems (Fig. 6.5). This system structure allows the EWSD switching system, for example (Fig. 6.7), to exercise decentralized control of the D-channel protocol for the ISDN basic access: the physical and logical functions of layer 1 and the complete layer 2 procedure (HDLC LAPD; see Sect. 4.3.4) for eight ISDN basic accesses are handled on the ISDN "Subscriber Line Module Digital" (SLMD) of the DLU, whereas control of layer 3 of the D channel protocol is undertaken in the "Line/Trunk Groups" LTG of the parent exchange.

6.2.5 Control

The switching functions of an ISDN local exchange may be roughly divided into the following functional areas, as shown in Table 6.2:

(1) Signaling
(2) Access control

(3) Switching between subscriber lines and trunks
(4) Interaction with the Operation and Maintenance functional unit.

The control functional unit comprises functional areas (2) through (4), and in practice these can be implemented as distributed functions. Thus, for instance, the EWSD system (Fig. 6.7) has a modular decentralized control structure to relieve the load on the coordination processor (CP) and this structure is characterized by a high degree of preprocessing in microprocessors of peripheral units – in particular within the group processors (GP) of the line/trunk groups (LTG). The specific control of a subscriber or trunk access line (2) is to a large extent performed independently by the appropriate group processor (GP), whereas switching (3) between the internal ports of different line trunk groups (LTG) is controlled by the coordination processor. For this purpose the line trunk groups, each of which is connected via an internal 8.192 Mbit/s time-division multiplex interface (124 channels each with 64 kbit/s) to the central switching network (SN) that operates at the same rate, use 64 kbit/s channels to exchange not only information between users but also control information with the coordination processor. The 64 kbit/s channels for the control information are switched through the switching network as semipermanent connections, with serial/parallel conversion between the serial 8.192 Mbit/s switching net-work interface and the parallel coordination processor interface being per-formed by the message buffer (MB). Through-connection of the 64 kbit/s user information channels via the switching network is controlled by the coordina-tion processor using setting commands to the switch group control (SGC).

6.2.6 Interexchange Signaling

The functional unit for interexchange signaling deals primarily with the CCITT common channel signaling system No.7 (shown by the abbreviation CCS in Figs. 6.2, 6.4 and 6.7). The system may be used for interworking both with the exchanges in the digital telephone network (Telephone User Part TUP) [6.16 through 6.20] and with other ISDN exchanges. For interexchange signaling in the ISDN, the message transfer part (MTP) already defined [6.41 through 6.48] of Signaling System No.7 can be used without further modification. To take account of the particular requirements of ISDN interexchange signaling, a new ISDN User Part (ISUP [6.22 through 6.26]) has been defined (see Sect. 6.3).

In addition, to take account of interworking between ISDN and existing analog parts of the telephone network, the appropriate channel-associated signaling (CAS) has to be handled in an ISDN exchange, too.

6.2.7 Operations, Administration and Maintenance

The functions for Operations, Administration and Maintenance (OAM or O&M) include the following functional areas [6.49]:

- Operation and administration functions associated with the creation, modification and expansion of system data, e.g. exchange, network, trunk group and subscriber access data.
- Cutting over and expanding exchange equipment and service features.
- Repair and maintenance functions to preserve the operability and secure the quality of service of the switching system.

The Q.500 series of Recs. also refers to OAM: The operations and maintenance *design objectives* applicable to local, combined, and transit exchanges are described in Rec. Q.542 [6.50] including network management controls, alarm handling and subscriber line maintenance and testing. The *interfaces* associated with operations, administration and maintenance (OAM) for local, combined and transit exchanges will be defined in Rec. Q.513 [6.51]. According to Rec. Q.513 there are two general classes of interfaces for OAM:

- Human-machine interfaces using the CCITT Man-Machine Language (MML) defined in the Z.300 series of CCITT Recs.
- Interfaces to OAM operations systems (OS) and to OAM work stations (WS). These will be based on the concept of the Telecommunications Management Network TMN (see below).

Operations and maintenance procedures and protocols associated with the *interexchange area*, i.e. with the common channel signaling network and the exchanges, are described in the Operations and Maintenance Application Part OMAP (see CCITT Rec. Q.795 [6.52] and Sect. 6.3).

OAM principles and functions for the *ISDN subscriber access*, e.g. fault tracing by means of loopbacks (see Sect. 4.1.2), are contained in the I.600 series of Recs. (cf. CCITT Rec. I.601 [6.53]). The Q.940 series of Recommendations [6.54] is intended to describe *user-network interface protocols* for the exchange of management information in accordance with the concept of the Telecommunications Management Network TMN (see below) and the Management Framework for Open Systems Interconnection.

The variety and complexity of today's telecommunication networks has led to a wide range of application-dependent and vendor-specific OAM solutions and management systems/protocols. From a network operator's point of view, however, a standardized OAM approach is preferable in order to allow for global control of all kinds of switching and transmission equipment in a network. This objective is reflected in the new concept of the *Telecommunications Management Network (TMN)* contained in the framework Recommendation M.30 [6.55].

According to Rec. M.30, a TMN (Fig. 6.8) provides an organized network structure to achieve the interconnection of various types of Operations Systems OS (i.e. network management centers) and telecommunication equipment (called Network Elements NE) using an agreed-upon architecture with standardized interfaces and protocols. The TMN can be viewed as a data communication network (DCN) independent of the telecommunication network it

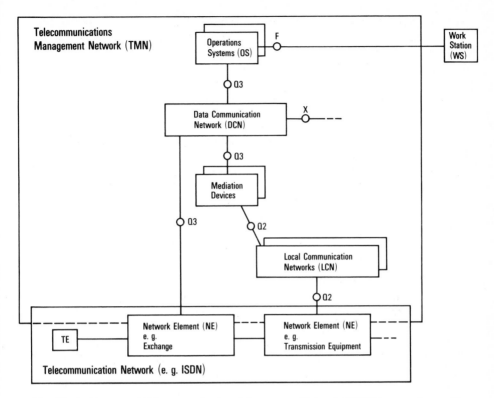

Fig. 6.8. Architecture of Telecommunications Management Network (TMN) in accordance with CCITT Recommendation M.30.
Q2   Interface for less complex network elements (e.g. transmission equipment) which is adapted to the Q3-interface via a mediation device
Q3   Interface supported by the operations system (OS). Q3 is used for connecting exchanges.
F   Interface for connecting a workstation (WS) for communication with the user
X   Interface to other networks (including other TMNs)
TE   Terminal equipment

controls. Via the DCN, the OS receives status information from and controls the operation of the components (NEs) of the telecommunication network (e.g. an ISDN). While exchanges are directly connected via the Q3-interface used by the OS, less complex or existing devices (e.g. transmission equipment) can be adapted to Q3 by means of suitable Mediation Devices (MD). More details on this future-oriented network management approach can be found in [6.56].

6.2.8 Timing and Network Synchronization

In synchronous digital networks the timing signals of the primary (2.048 Mbit/s) or secondary (8.448 Mbit/s) multiplex signals produced by the individual exchanges must match very precisely in order to avoid corruption of the user information (see Sect. 7.6). The *central clock generator* (*CCG*) functional unit (cf. Figs. 6.5 and 6.8) has the task of matching the timing signal generated in the exchange to the reference frequency fed in from an external source – e.g. from a reference clock located centrally in the network – and to distribute the resultant timing information within the exchange in such a way that the synchronization of the 64 kbit/s time slots is maintained as they pass through the exchange.

6.2.9 Interworking and Access to Special Equipment

As already explained in Sect. 6.1, specialized equipment for certain additional functions is not provided in every ISDN exchange. Such equipment includes

– service modules performing high layer functions (vendor feature nodes VFN),
– equipment for interworking with dedicated data networks,
– network data bases (service control points SCP, see Sect. 6.3.7),
– equipment for establishing leased line networks (digital cross connects DCC).

For reasons of economy and practical implementation, these functions are centralized. Additional information on the provision of packet switched services to ISDN subscribers as well as on interworking between ISDN and dedicated data networks can be found in Sect. 4.4

Unlike centralized interworking with dedicated data networks, interworking with the conventional telephone network is effected in all ISDN exchanges (cf. Sect. 6.2.2). The significance of this for implementation is that the analog and digital interexchange interfaces and signaling procedures of the telephone network have to be present in every ISDN exchange (Figs. 6.4 and 6.7).

6.3 Interexchange Signaling in the ISDN

6.3.1 Basic Characteristics of Interexchange Signaling Using CCITT Signaling System No. 7

To set up and clear down 64 kbit/s circuit switched network connections and control ISDN services and supplementary services, the ISDN exchanges concerned must be able to exchange signaling information with one another. Interexchange signaling in the ISDN employs CCITT Signaling System No. 7. This section deals with the basic features of this system [6.12 to 6.14, 6.57] insofar as they provide an understanding of the ISDN-specific additions and extensions to the signaling system.

Unlike conventional signaling systems for telephone networks, e.g. CCITT Signaling Systems No. 4 and No. 5 [6.58, 6.59], CCITT No. 7 is a *common channel signaling system* (Fig. 6.9.a). Common channel signaling differs from the signaling systems mentioned above in that the signaling information which relates to the 64 kbit/s user information channels is transferred in separate

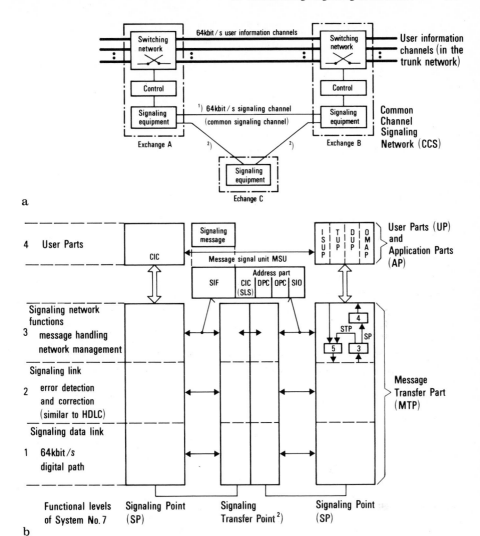

Fig. 6.9a, b. Common Channel Signaling. **a** Principle; **b** Protocol Architecture of CCITT Signaling System No.7.

CCS	Common Channel Signaling	OPC	Originating Point Code
ISUP	ISDN User Part	DPC	Destination Point Code
TUP	Telephone User Part	SIO	Service Information Octet
DUP	Data User part	SI	Service Indicator
OMAP	Operations and Maintenance Application Part	SP	Signaling Point: source or sink (with processing) of signaling messages
MSU	Message Signal Unit		
SIF	Signaling Information Field	STP	Signaling Transfer Point: Transfer of signaling messages without processing
CIC	Circuit Identification Field		
SLS	Signaling Link Selection		

[1] Associated signaling mode
[2] quasi-associated signaling mode
[3] Message discrimination
[4] Message distribution
[5] Message routing

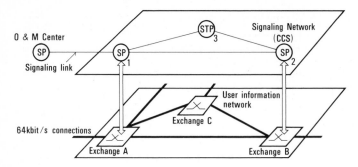

Fig. 6.10. Separate Networks for User Information Transfer and Signaling.

CCS	Common Channel Signaling
O&M	Operations and Maintenance
SP	Signaling Point
STP	Signaling Transfer Point
Signaling modes:	1–2 associated
	1–3–2 quasi-associated
A–B	Signaling relation

64 kbit/s signaling channels which are shared by a number of user information channels. The signaling channels between the exchanges together constitute an independent signaling network (common channel signaling network) based on message interleaving which is completely separate from the user information network (Fig. 6.10).

The advantages of common channel signaling compared to channel-associated signaling include:

- *Signaling simultaneous with transfer of user information* is possible with network connection already set up.
- *Short call setup time* as a result of high capacity (64 kbit/s) of the signaling channels.
- *Virtually unlimited set of signaling elements* (messages, message parameters) in conjunction with
- *Great flexibility* with respect to new requirements (such as those of ISDN), including the introduction of new services and supplementary services.
- The structure of signaling elements is particularly suitable for *processing in SPC exchanges.*
- *Less expensive than a channel-associated procedure* since centralized signaling equipment of a single 64 kbit/s signaling channel can be used to simultaneously control approximately 1000 user information channels.
- *Possibility of use of the signaling network for applications outside signaling,* e.g. for transfer of operations and maintenance information between the exchanges and O&M centers.
- *Secure transfer* of the signaling elements (i.e. safeguarding against transmission errors).

Interaction between signaling network and user information network occurs during the link-by-link setup of a 64 kbit/s circuit switched network connection only in those exchanges in which the signaling information is passed to the control entity for processing (cf. Sect. 6.2.5). The latter then in turn performs the switching network settings (Fig. 6.9a). Accordingly, those nodes of the signaling network in which there is *application-oriented* processing (or creation) of signaling messages – i.e. the relevant end points (sources, sinks) of a signaling relation between two adjacent exchanges – are referred to as *Signaling Points* (SP see Fig. 6.10). In contrast, the *Signaling Transfer Points* (STP) merely perform the *transport-oriented* signaling functions, i.e. they direct the incoming signaling messages to a destination signaling point without any further processing (routing function of the signaling network). In practice SP and STP functions occur in combination in the same exchange.

The explicit separation of transport-oriented and application-oriented functions described here is reflected in the protocol architecture of CCITT Signaling System No.7. As shown in Fig. 6.9b, the description of System No.7 is divided into:

- *A Message Transfer Part (MTP) common to all applications*, and
- *Separate User Parts e.g. for telephony (Telephone User Part TUP* [6.16 to 6.20]), *circuit-switched data services (Data User Part DUP* [6.21] for common channel signaling in data networks) and application parts e.g. for *operations and maintenance tasks (Operations and Maintenance Application Part OMAP* [6.52].

The special requirements of ISDN interexchange signaling result from service integration – *one* common user part for telephony *and* all non-voice services is needed – and from the control of innovative ISDN supplementary services. Accordingly, a separate ISDN User Part (ISUP) has been created (cf. [6.15, 6.22 through 6.27] and Sects. 6.3.3 through 6.3.6).

6.3.2 The Message Transfer Part (MTP)

The following paragraphs briefly describe the main MTP functions in SS No.7 levels 1 through 3. *Level 1* comprises the transmission and access functions of a *physical signaling channel* (signaling data link), while *level 2* ensures *reliable transfer* with protection against transmission errors of the message signal units (messages) over a signaling link to the next node of the signaling network (signaling point SP or signaling transfer point STP) using a procedure similar to HDLC.

Level 3 mainly contains the signaling network functions necessary for *signaling message handling*, i.e. the functions for the distribution of the messages received at their destination point to the correct user part within the exchange, or for routing of messages destined for another exchange to the correct outgoing signaling link. Level 3 also accommodates the overall control of operations and maintenance functions for the signaling network (*signaling network management*); this includes functions such as distributing the signaling traffic load to a number of signaling links and reconfiguration of signaling routes in case of failures.

Control information used for level 3 functions includes the address of the originating exchange (Originating Point Code OPC), the address of the destination exchange (Destination Point Code DPC), identification of the signaling link chosen (Signaling Link Selection SLS) and the service information octet (SIO). Using the Service Indicator (SI), which is a subfield of the SIO, an incoming message is routed to the correct user part (ISUP, TUP...) within the same exchange (*message distribution function*), provided the preceding evaluation of the DPC indicated that this message was actually destined for that particular signaling point (*message discrimination function*). The transfer messages from other exchanges on incoming signaling links as well as the messages from the user parts within the exchange destined for onward transmission are *routed* to the appropriate outgoing signaling link in the destination direction on the basis of the DPC (connectionless datagram principle).

As a consequence of the shared usage of the common signaling channel by a number of network connections, the allocation of each message to a particular user information channel (circuit) has to be uniquely established; this is ensured by the CIC (Circuit Identification Code). Since the signaling link selection code (SLS) mentioned previously is a component of the CIC, this ensures that all messages associated with *one* signaling link take the same path through the common channel signaling network (in-sequence transfer of signaling messages).

6.3.3 Signaling Relations Between ISDN Exchanges

When the ISDN is introduced the circuit-related type of signaling described in Sect. 6.3.2 between pairs of adjacent exchanges throughout the network connection path will no longer be sufficient. This conventional *link-by-link* signaling between adjacent signaling points which is related to the set-up and release of a circuit between the respective exchanges, has been supplemented for the ISDN user part (ISUP) by the new function of *end-to-end* signaling between originating and terminating ISDN local exchanges e.g. between signaling points SP_A and SP_B in Fig. 6.11a. Even though it may be related to a call, end-to-end signaling is not directly related to the control of the respective circuit switched connection. Thus, end-to-end signaling can be viewed as the capability to transfer signaling information directly between the endpoints of a circuit-switched connection (transfer of call-related information or of user-to-user information). In addition it may even be employed for exchanging signaling information between signaling points which are not interconnected by a circuit-switched connection (see Sect. 6.3.7).

In case of end-to-end signaling, the intermediate signaling points of the transit exchanges (cf. SP_T in Fig. 6.11a) are by-passed. This can be achieved by

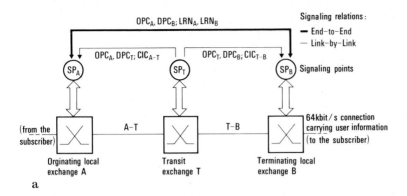

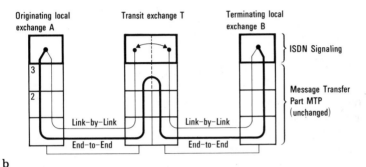

Fig. 6.11a, b. Signaling Relations Between ISDN Exchanges. **a** Principle; **b** Layer structure.

OPC	Originating Point Code
DPC	Destination Point Code
CIC	Circuit Identification Code (only in case of link-by-link signaling)
LRN_A, LRN_B	Local reference numbers in exchanges A or B to identify end-to-end signaling relations (logical connections) between exchanges A and B

the SCCP method described in more detail in Sect. 6.3.4, where the transit exchanges act as signaling transfer points STP (Fig. 6.11b). ISDN exchanges providing interworking with the telephone network and gateway exchanges to foreign ISDNs or to data networks act as originating and terminating exchanges as regards end-to-end signaling.

The main objectives of end-to-end signaling are

- *Relieving the load on the ISUP in the transit exchanges* so that they do not have to process the extra signaling traffic resulting from the control of supplementary services, such as changing from voice to data and back during an established call (see in-call modification in Sect. 6.3.6). These supplementary services are always controlled in the ISDN local exchanges.
- The option of *signaling when a network connection does not exist* or it has already been cleared down, e.g. signaling for a supplementary service such as *completion of calls to busy subscribers.*

The *end-to-end signaling* functions include messages for

- Requesting a supplementary service:

  FRQ   Facility Request
  FACD   Facility Accepted
  FRJ   Facility Reject

- Transmitting information relevant to supplementary services:

  FIN   Facility Information

- Deactivation of a supplementary service:

  FDE   Facility deactivated

- Information of the called user when the calling user suspends the call (without releasing it) and then resumes it again e.g. for moving a terminal from one socket to another in the case of a bus configuration (see Sect. 4.3.5.4):

  PAU   Pause
  RES   Resume

- Information request from an interworking exchange or gateway exchange

  IRM   Information request
  INF   Information

- Transfer of user-to-user information (cf. Fig. 6.18); the function of the ISUP message USER INFO corresponds to the D-channel message with the same name (see Sects. 4.3.5.5 and 6.4.3):

  USER INFO User-to-user information

The setup and release of the circuit switched connection (basic call control), on the other hand, is controlled by *link-by-link* messages, as it is in the TUP. These messages can be evaluated by each exchange involved in the path of the overall network connection. A significant difference from the TUP is that call release can be initiated by the called or the calling user, as in data networks. The category of link-by-link transferred ISDN signaling messages includes:

- *Call establishment* messages (cf. Fig. 6.13a):
  IAM   Initial address

  Initializing message sent in the forward direction to initiate seizure of an outgoing circuit and to transmit address information (complete directory number of the called user for en bloc dialing or for overlap (digit-by-digit) dialing insofar as required for routing to the national destination exchange); other information required for call processing, e.g. supplementary services to be taken into account during call setup (e.g. reverse charging).

  SAM   Subsequent address

  Transports the digits not contained in the IAM if necessary (e.g. in case of overlap dialing).

ACM   Address complete
        The called user is free (and compatible) and has responded to the call with ALERTing.

ANS   Answer
        Call accepted by the called user with CONNect.

- *Release* messages (cf. Fig. 6.13b)

REL   Release
        With REL, one (or both) of the local exchanges initiates the release of the user information channel in the network, as soon as the local subscriber releases with DISConnect. Receipt of a REL message causes the transit exchanges or the released local exchange to disconnect the user information channel and to send an RLC message (see below) in the backward direction as an acknowledgement.

RLC   Release Complete
        Acknowledgement for REL.

6.3.4 Protocol Architecture of ISDN Interexchange Signaling

To implement the end-to-end signaling between ISDN originating and terminating local exchanges explained in the previous section, the transport-oriented functions implemented by the message transfer part MTP of Signaling System No. 7 must be expanded. In order to avoid affecting the existing user parts which are based directly on the MTP (TUP . . .), this functional expansion has been undertaken above the otherwise unaltered MTP (Fig. 6.12).

The ISUP signaling procedures in CCITT Rec. Q.764 [6.25] make provision for two different methods for end-to-end signaling:

- With the *SCCP method* the ISDN user part ISUP [6.22 through 6.26] makes use of the services of a new *Signaling Connection Control Part* (SCCP, see CCITT Rec. Q.711 through Q.714 [6.60 through 6.63]) introduced between the Message Transfer Part and the ISUP. In accordance with the principle of protocol layering this means that ISUP messages for end-to-end signaling are conveyed by SCCP messages (see Data Form 1 below) which in turn are transported in the signaling information field (SIF) of MTP message signal units (MSU). In contrast to the pass-along method (see below), the SCCP method is independent of the presence of a circuit-switched connection between the message originating and terminating local exchanges. The end-to-end signaling connections between the ISDN local exchanges are established in this case exclusively via the MTP function in the transit exchanges (see Fig. 6.11).

- In the case of the *pass-along method* the end-to-end signaling information is passed on a link-by-link basis along the path of transit exchanges at which the sections of the circuit-switched connection are interconnected. Forwarding of

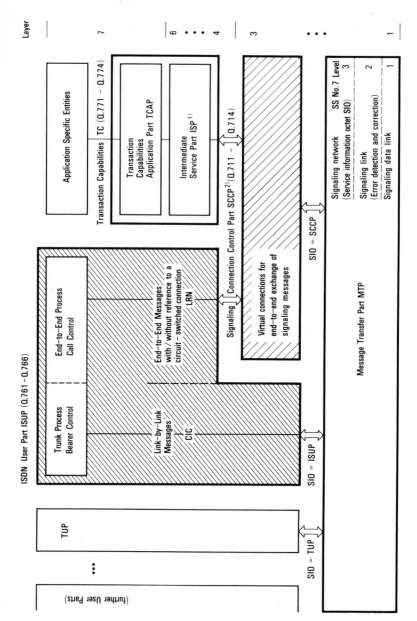

Layer

ISDN User Part ISUP (Q.761–Q.766)

Application Specific Entities

Transaction Capabilities TC (Q.771 – Q.774)

Transaction Capabilities Application Part TCAP

Intermediate Service Part ISP [1]

Connection Control Part SCCP[2] (Q.711 – Q.714)

Virtual connections for end-to-end exchange of signaling messages

Signaling

Signaling network (Service information octet SIO)   SS No. 7 Level   3

Signaling link (Error detection and correction)   2

Signaling data link   1

Message Transfer Part MTP

End-to-End Process Call Control

End-to-End Messages with / without reference to a circuit – switched connection LRN

Trunk Process Bearer Control

Link–by–Link Messages CIC

TUP

(further User Parts)

SIO = SCCP

SIO = ISUP

SIO = TUP

Application-oriented SS No. 7 functions for ISDN

Enhancement of transport–oriented MTP functions

incoming end-to-end messages to an outgoing link to the next exchange is effected by the ISUP in the transit exchanges as in the case of link-by-link messages (see Fig. 6.11). End-to-end messages are not processed in the transit exchanges, however, but simply readdressed for subsequent transmission on the outgoing link.

With the SCCP method, the ISUP has two separate layer interfaces for transporting signaling messages, as shown in Fig. 6.12.:

- a direct *interface to the MTP* for *link-by-link* messages and
- an indirect access to the MTP via the *interface to the SCCP* for *end-to-end* messages.

As illustrated in Fig. 6.12, it is the task of the SCCP to raise the level of MTP services up to that available at the OSI layer 3 interface. The OSI network layer service is thus provided by the combination of MTP plus SCCP, also referred to as the Network Service Part NSP (cf. Fig. 6.16). As a consequence, non-circuit related S.S. No. 7 signaling functions built on top of the SCCP in accordance with the OSI reference model (CCITT Rec. X.200) are called application parts such as the Operations and Maintenance Application Part OMAP [6.52] and the Transaction Capabilities Application Part TCAP (CCITT Recs. Q.771 through Q.774 [6.64 through 6.67]). The TCAP is used to access network data bases (see Sects. 6.1 and 6.3.7), e.g. for number translation in conjunction with the freephone service (e.g. service "800" or "130").

End-to-end signaling connections can be set up and released by means of the following SCCP messages (cf. Fig. 6.13):

CR   Connection Request
CC   Connection Confirm
RLSD   Release
RLS   Release Complete

The SCCP message

DT1   Data Form 1 (cf. Fig. 6.20)

Fig. 6.12. Protocol Architecture of CCITT Signaling System No. 7.
ISUP   ISDN User Part: ISDN interexchange signaling as per CCITT Rec. Q.761–766 [6.22 through 6.27]
TUP   Telephone User Part
SCCP   Signaling Connection Control Part as per CCITT Rec. Q.711–714 [6.60 through 6.63]
TC   Transaction Capabilities as per CCITT Rec. Q.771–774 [6.58 through 6.61] provide functions and protocols for a large variety of applications distributed over exchanges and specialized centers in telecommunications networks
TCAP   Transaction Capabilities Application Part [6.64 through 6.67]
CIC   Circuit Identification Code
LRN   Local Reference Number for identifying end-to-end signaling connections
SIO   Service Information Octet
[1] comprises OSI Transport, Session and Presentation Layers (X.224, X.225)
[2] provides OSI Network Layer Service in conjunction with the MTP: connectionless (classes 0,1) or connection-oriented (classes 2 to 4)

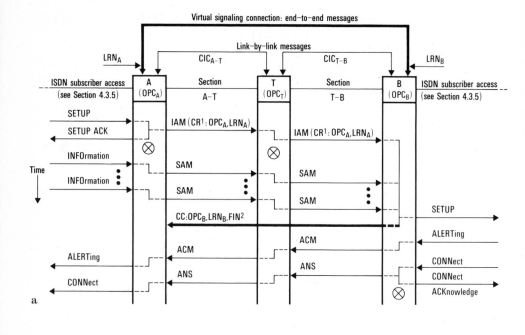

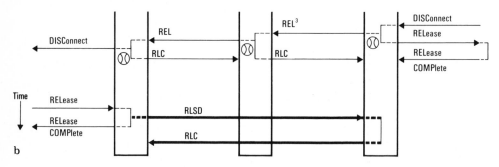

Fig. 6.13a, b. Set-up and Release of a Circuit Switched Connection in the ISDN. **a** Set-up of the 64 kbit/s connection (with overlap sending) and of the virtual end-to-end signaling connection; **b** Cleardown of the 64 kbit/s connection (initiated by the called terminal) and of the virtual end-to-end signaling connection. IAM, SAM, ACM, ... ISUP messages to set up and release the 64 kbit/s connection (see Sect. 6.3.3).
CR, CC, RLSD, RLC: SCCP Messages to set up and release the virtual signaling connection between local exchanges A and B (see Sect. 6.3.4).

A   Originating local exchange
T   Transit exchange
B   Terminating local exchange
OPC   Originating Point Code
DPC   Destination Point Code
CIC   Circuit Identification Code
LRN   Local Reference Number for identifying the end-to-end signaling connection
[1] Implicit setup of end-to-end signaling connection by embedding CR in IAM
[2] Facility information (FIN) message: A-exchange is informed on ISDN subscriber-related service attributes available at B-subscriber
[3] Release can be initiated either by the calling party (A-subscriber) or the called party (B-subscriber)

is used as a transparent transport container for end-to-end signaling messages of the ISUP via existing end-to-end signaling connections.

As already mentioned in Sect. 6.1, two principal functions of an ISDN exchange are call control and bearer control (i.e. connection control). In the updated CCITT Recs. Q.761 to Q.766 to be published in the Blue Book (1989), call control functions and bearer control functions common to all bearer services will be contained in a single functional block comprising both end-to-end and trunk processes (Fig. 6.12). As compared to discussions in CCITT aiming at a *separated* ISUP consisting of a functional block for call control and separate specialized functional blocks for each bearer service, the approach in the Blue Book and in the Red Book is referred to as the *monolithic* ISUP:

- The *end-to-end process* performs control and coordination of requests for circuit switched connections, packet switched connections (in case of single step call establishment, see Sect. 4.5), virtual signaling connections (SCCP), etc.
- The *trunk process* establishes various types of network connections on a link-by-link basis; in contrast to the end-to-end process it is therefore present in the transit exchanges, too.

6.3.5 Implementation of ISDN Interexchange Signaling in the Exchange

With the Siemens EWSD system, line-type-specific control of subscriber and trunk lines is provided largely by the group processors (GP) in the peripheral line/trunk groups (LTG), in keeping with the concept of a modular, decentralized control structure (Sect. 6.2.5 and [6.4, 6.6]). The *application-oriented No. 7 functions* of a signaling point SP, i.e. those of the ISDN User Part ISUP for example, are thus implemented for all user information channels routed via a particular LTG in the GP of the LTG concerned (Fig. 6.7).

The LTG "C" units on the interexchange side assume the link-by-link signaling functions under these circumstances, whereas the end-to-end signaling is implemented in the LTG "B" unit on the subscriber access side on account of the close connection with user-network signaling. This also applies to Signaling Connection Control Part SCCP which is another "user" of the message transfer part MTP and expands the MTP functions to include end-to-end transport of ISUP messages.

The No. 7 functional levels 2 (signaling link) and 3 (signaling network) of the *Message Transfer Part MTP*, i.e. functions such as those of a signaling transfer point STP, are handled by a Common Channel Signaling Network Control (CCNC) subsystem designed specially for the common channel signaling network. The 64 kbit/s signaling channels are routed transparently from the LTG "C" units via fixed switching network paths (semipermanent connections) to the CCNC. Incoming ISUP messages intended for a signaling point SP within the same exchange are directed by the CCNC via the CCNC/CP interface and the semipermanent 64 kbit/s control channel between the message buffer MB and the GP (cf. Sect. 6.2.5) to the correct destination LTG for evaluation; outgoing ISUP messages from the local signaling point SP go in the opposite direction. The CCNC undertakes routing of transfer messages as well as outgoing messages from the local SP to a LTG C corresponding to the outgoing signaling link.

6.3.6 Intelligent Network

6.3.6.1 Flexible Service Control Architecture

According to the new service control concept of the *Intelligent Network (IN)* [6.29, 6.32 through 6.34], as already mentioned in Sect. 6.1, control of calls requiring special handling can be distributed between the ISDN exchanges, called *Service Switching Points (SSP)*, on the one hand and internal network data bases, termed *Service Control Points (SCP)* on the other (Fig. 6.14). When

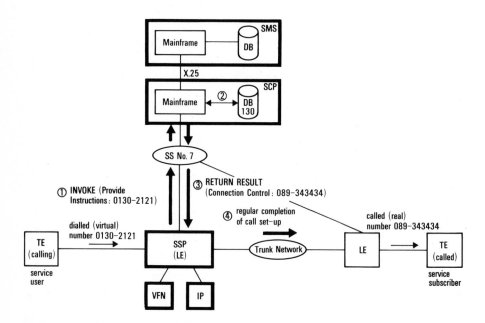

Fig. 6.14. Intelligent Network (e.g. Freephone Service)
SMS   Service Management System
SCP   Service Control Point (network data base)
SSP   Service Switching Point (ISDN exchange)
TE   Terminal Equipment
VFN   Vendor Feature Node
IP   Intelligent Peripheral
DB   Data Base
LE   Local Exchange
INVOKE,   TCAP messages (see Sect. 6.3.6.2)
RETURN RESULT
Provide Instructions,   Functional Components
Connection Control

performing call establishment for IN-services, the SSP accesses appropriate service information (control logic and/or service data) stored in the SCP via the Transaction Capabilities of the Signaling System No. 7 (see Sect. 6.3.6.2).

The overall objective of IN is improved flexibility of service control in order to enable network operators to introduce new services and service features more rapidly, even on a trial basis, merely by *SCP modifications which are fully transparent to the network exchanges* (i.e. the SSPs). The IN concept also allows for service customization to cater for unique customer needs.

Another important aspect is to give customers control over their service features by allowing direct customer access to the *Service Management System* (*SMS*), e.g. for changing the call routing information in conjunction with the freephone service (service 800 see below) or even to reconfigure a virtual private network embedded in the resources of the public ISDN.

By means of suitable triggers in the switching software, the SSP finds out whether a given call is to be handled by the SCP or whether it can be completed by the SSP. In the case of the freephone service (Fig. 6.14), i.e. when the call is paid by the called service subscriber, a special access code (e.g. 800 or 130) contained in the dialed number is used as a trigger. This causes the SSP to access the SCP for conversion of the dialed *virtual* number (0130-2121) into a valid *real* number (089-343434) of the called party. By implementing the freephone service within IN it gains in flexibility, since the service subscriber may change the called real number by accessing the SMS.

6.3.6.2 Transaction Capabilities via CCITT Signaling System No. 7

Prerequisites for distributed IN service control as described in Sect. 6.3.6.1 are the availability of CCS No. 7 as well as protocols for *non-circuit-related interactions* via CCS No. 7, e.g. query/response type interactions. Towards this end, *Transaction Capabilities TC* have been defined in CCITT Recommendations Q.771–774 [6.64 through 6.67]. As shown in Fig. 6.12, TC are built on top of the Signaling Connection Control Part SCCP. The TC comprise the *Transaction Capabilities Application Part TCAP* in Layer 7 and supporting standard OSI-protocols in Layers 4 to 6.

In this way, ISDN provides an optimum framework for the introduction of IN services. Even though not dependently linked, IN and ISDN enhance one another because they provide complementary capabilities with respect to services, i.e. flexible service control on the one hand and powerful signaling capabilities on the other.

6.4 Corporate ISDN Networks

6.4.1 Fundamental Solutions

The advantages of integrated voice and data features inherent in the ISDN concept are most useful for business applications. As a consequence, ISDNs can be expected to form the "hub" of current and future business communication and office automation networks.

On the other hand, corporate networks linking multiple company locations have traditionally been characterized by enhanced requirements as compared to public networks. Examples include:

- Wider range of features;
- Private facilities, e.g. tie trunks, paging, recorded dictation;
- Private numbering plan;
- Attendant features, e.g. attendant-assisted calls, directory assistance, message desk service;
- Interconnection of multiple customer locations with feature transparency, as if the extensions were on a single switch (e.g. city-wide Centrex);
- Communication management features, e.g. station message detail recording (such as reports on dates, times and length of calls), traffic data to customer premises equipment;
- Customer control features, e.g. access to outgoing facilities, customer station rearrangement (with regard to feature and number allocation).

To meet the extended communication requirements of corporate ISDNs, fundamentally different solutions can be utilized. Depending on the regulatory

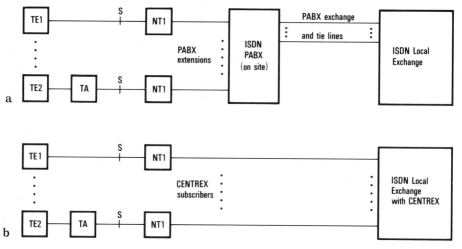

Fig. 6.15a, b. Basic Solutions for ISDN Corporate Networks. **a** On-site ISDN PABX (private corporate network); **b** Centrex system (corporate network utilizing the local exchange for PABX-like features).

PABX   Private automatic branch exchange
CENTREX Central exchange
TE1   ISDN terminal equipment with S/T interface
TE2   Terminal with non-ISDN interface
TA   Terminal adaptor (see Sect. 4.4)
NT1   Network termination

framework and the customer's requirements, these different solutions can in practice complement each other rather than substitute for each other:

- *Private corporate networks* composed of interconnected on-site ISDN private automatic branch exchanges (ISDN PABX or ISPBX) which provide the required features such as those listed above (refer to Fig. 6.15a and [6.5, 6.68, 6.69]); this topic is covered in Sects. 6.4.2 and 6.4.3.
- Corporate networks utilizing the local exchange for providing the required features (such as those listed above). This approach is traditionally known as *Centrex (Central Exchange)* where local public ISDN exchanges provide PABX-like features (refer to Fig. 6.15b and [6.70, 6.71]). In practice, a Centrex group may be composed of both PABX and directly connected subscribers. Centrex can provide for both dedicated and virtual private networks. The distinction is that virtual private networks utilize the public trunk network with a private software-defined overlay guaranteeing a specific amount of customer access and control; dedicated private networks, on the other hand, employ private trunks.

6.4.2 Structure and Features of an ISDN PABX

Figure 6.16 shows the structure of an ISDN PABX, which forms the central element of an integrated office communication system (e.g. Siemens System HICOM [6.5, 6.68]. Such a system may comprise the following functional areas:

- Basic circuit-switched system,
- Local area network (LAN),
- Packet Handler providing X.25 data communication services,
- Interconnection of PABXs to form ISDN corporate networks (see Fig. 6.18 and Sect. 6.4.3),
- Special equipment for high layer services and functions beyond information transport, namely
  - *Storage*: Private database systems, inhouse Videotex centers
  - *Processing*: Data processing systems (DPS)
  - subscriber-like connection of a DPS to allow local or remote terminals to access the DPS via the PABX
  - Connection of a DPS via an administration and data server for operations and maintenance linkage between PABX (control) and DPS, e.g. for recording call charge data, centralized operations, remote operations administration and maintenance, and DPS-controlled call setup.
  - *Store and forward communication*: Voice mailbox and text mailbox systems (message handling systems)
  - *Compatibility*: protocol conversion between terminal and host computer, interworking between different services (e.g. from teletex to facsimile).

The structure depicted in Fig. 6.16 is based on a digital time-division multiplex switching system; this implicitly defines the following characteristics for a

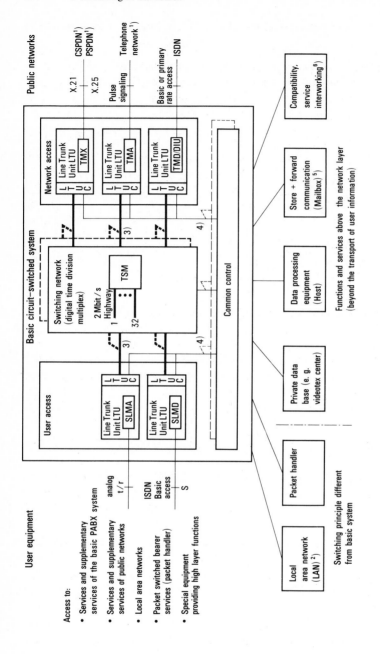

corporate network for office communications:

- *One* network for *all* types of communication, including voice communication,
- *Star* topology with *central* control, i.e., the network structure currently widely used in PABX networks is retained, and the use of the existing inhouse infrastructure of the telephone line network is continued;
- *Circuit switching* supplemented by access to X.25 packet switched services.

The *basic circuit-switched system* (Fig. 6.18) in the ISDN PABX is formed by a digital time-division multiplex switching system. Structure and functions of a private ISDN switching system thus broadly correspond to those described in detail in Fig. 6.4 for public ISDN exchanges. In addition to the peripheral units for user access and for network access, which perform the interface and signaling functions, the PABX, like the public exchange, has central components, duplicated for security reasons, i.e. a non-blocking digital switching network using time-division multiplexing for through-connection of the 64 kbit/s channels and an SPC common control. The functional units (Line Trunk Units LTU) of the periphery accommodate modules for the user access (Subscriber Line Module SLM. . .) or for the network access (Trunk Module TM. . .). Typically, the LTUs are linked to the switching network via internal time-division multiplexed 2 Mbit/s highways. Control information is exchanged with common control via separate control paths to relieve the load on the switching network.

In the *user access area* of the ISDN PABX (Fig. 6.17a) the internationally standardized D-channel protocol (S interface) for public ISDN systems can also be used, at least as a basis. The resulting advantage of terminal portability between public and private subscriber lines does not have to be at the cost of reduced flexibility – e.g. with regard to the larger range of features and

---

Fig. 6.16. Basic Structure of an ISDN PABX as the "Hub" of an Office Communication System.

LTU	Line trunk unit
LTUC	LTU control
SLMA	Subscriber line module analog
SLMD	Subscriber line module digital
TSM	Time stage module
TMX	Trunk module with X. interface signaling
TMA	Trunk module analog
TMD	Trunk module digital (ISDN basic access)
DIU	Digital interface unit (ISDN primary rate access)
CSPDN	Circuit-switched public data network
PSPDN	Packet-switched public data network

[1] as an interim solution prior to the introduction of a public ISDN that provides interworking with dedicated data networks and with the telephone network

[2] Self-contained network with separate numbering scheme and with non-ISDN compatible access interfaces and protocols

[3] Exchange of user information via a 2 Mbit/s highway (internal time division multiplex system)

[4] Exchange of signaling information (control information)

[5] Servers for voice mail and text mail and store-and-forward features for facsimile and teletex terminals

[6] Protocol conversion between terminals and data processing equipment (DPE), interworking between common control and DPE or interworking between teletex and facsimile terminals

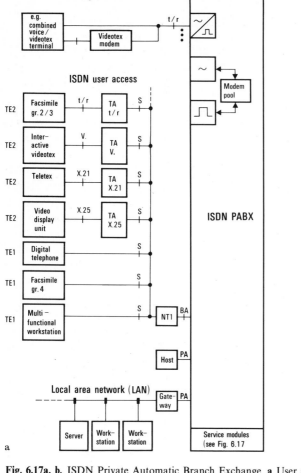

Fig. 6.17a, b. ISDN Private Automatic Branch Exchange. **a** User access; **b** Access to public and private networks.

TE1	ISDN terminal equipment with S interface
TE2	Terminal with non-ISDN interface
TA	Terminal adaptor (see Sect. 4.4)
NT1	Network termination
BA	ISDN basic access: $B + B + D_{16}$
PA	ISDN primary rate access: $30 \times B + D_{64}$ (2.048 Mbit/s) or $23 \times B + D_{64}$ (1.544 Mbit/s)
PH	Packet Handler (see Sect. 4.4.4)
IWU	Interworking unit (see Sect. 4.4)
DTE	Data terminal equipment
ISUP	ISDN interexchange signaling (ISDN User Part, see Sect. 6.3.3 to 6.3.5)
PABX	Private automatic branch exchange
PSPDN	Packet-switched public data network
CSPDN	Circuit-switched public data network
PSTN	Public switched telephone network

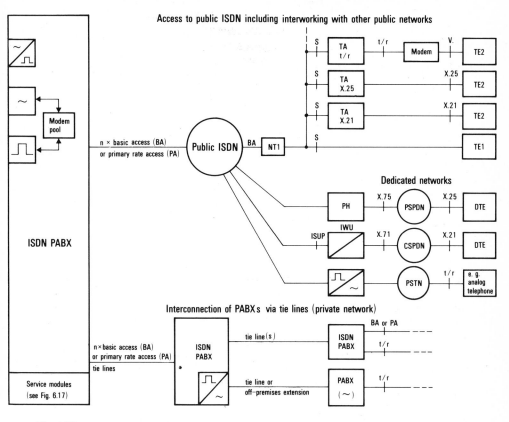

Fig. 6.17b

supplementary services traditionally provided by the PABX as compared to public networks – since in the standard CCITT signaling protocol provision has already been made for the protocol to be tailored to the needs of PABXs, i.e. by additional information elements in the signaling messages (cf. Sect. 4.3.5).

When ISDN PABXs are introduced, it must be ensured that the *existing analog terminal environment* can remain connected, at least for a transitional period: when analog subscriber lines are connected *directly* to a t/r interface, analog/digital conversion of the user information and conversion of the signaling for call establishment and clearing must be performed in peripheral units of the PABX – just as in the case of analog subscriber lines connected to public digital exchanges. For existing non-voice terminals of the analog telephone network it is an advantage to have an *indirect* connection via the ISDN user access with a terminal adaptor (TA) converting either the analog t/r interface (TA–t/r) or the digital V. interface (TA–V.) to the S interface (cf. Fig. 6.17a). In contrast to direct connection, indirect connection allows ISDN supplementary services to be implemented via the S interface using the terminal adaptor.

Even before the corresponding services are offered in a public ISDN or before the necessary interworking facilities are implemented from the public ISDN to the existing dedicated data networks (Fig. 6.17b), an ISDN PABX may provide the following network connections (Fig. 6.16) on the *network side* in addition to the connection to the ISDN:

- to the conventional telephone network,
- to a circuit-switched public data network (CSPDN),
- to a packet-switched public data network (PSPDN).

ISDN PABXs with direct dialing-in capabilities can be regarded as switching equipment similar to the exchanges of the public ISDN since they handle services and supplementary services for their users as the public ISDN local exchange does for public ISDN subscribers. However, the ISDN *inter-exchange signaling protocols* of CCITT Signaling System No. 7 (Message Transfer Part, ISDN User Part ISUP) described in Sect. 6.3 are not used for interfacing ISDN local exchanges, but rather the D-channel protocols designed for the subscriber access side of the public ISDN (see Sect. 4.3); this also applies to the primary rate access. This produces uniform signaling protocols for the user and network access side of the ISDN PABX, provided that on the user access side of the ISDN PABX, as already discussed, the CCITT standard D-channel protocol of the public ISDN is applied (cf. Sect. 6.4.3).

The basic circuit-switched system can be supplemented by *higher layer functions*, i.e. functions which go beyond pure information transport (Fig. 6.16). This extension of the basic system by storage, processing, store-and-forward communication and compatibility capabilities is particularly cost-effective when implemented by modular integration in the form of specialized equipment, commonly referred to as *service modules* or *servers*. Service modules (for voice and text mail, for example; cf. Sect. 2.1) can be accessed just like private data base systems or data processing systems for local or remote workstations by being connected to the PABX in the same manner as subscribers or other PABXs. They are also connected internally to central control for O&M purposes.

6.4.3 Tie Line Traffic Between ISDN PABXs

Figure 6.18 gives an overview of the networking arrangements and of the interface categories used with PABXs in an ISDN environment. Interface types include interfaces to *user equipment*, interfaces to the *public network* and interfaces to *other PABXs*. As already mentioned in the previous section, ISDN PABXs use, unlike the public ISDN exchanges, the same type of protocol for user access as well as for inter-exchange signaling, i.e. the D-channel protocol defined for public ISDNs (see Sect. 4.3) plus some "upward compatible" enhancements specific to PABX application.

Thus, the overall scope of application of the D-channel signaling protocol in the PABX environment includes the following *signaling relations of an ISDN PABX* (see Fig. 6.18a):

(1) Signaling between terminal and ISDN PABX (extension to PABX),
(2) Signaling to set up exchange line (EL) connections (PABX to ISDN local exchange),
(3) Signaling for tie line (TL) connections (PABX to PABX, see reference point Q).

ISDN PABXs, in a basically similar fashion to analog PABXs, make use of *tie lines* TL (for the basic traffic load) and exchange lines EL (essentially for peak traffic load) to handle tie line traffic between PABXs (Fig. 6.18). For tie line traffic between ISDN PABXs via the public network the ISDN related new supplementary services can also be employed on a network-wide basis, i.e. between users connected to different ISDN PABXs without restrictions; this is because in the ISDN signaling information can be exchanged between the ISDN PABXs involved even if there is an active call via a tie line in existence (cf. user-to-user signaling in Sect. 4.3.5.5). This is shown in Fig. 6.18a using the example of *change of service* for an established call from telephony (communication between terminals TE-A1 and TE-B1) to non-voice communication (e.g. facsimile communication between terminals TE-A2 and TE-B2).

Similar considerations apply to the network side of ISDN PABXs. Here the ISDN D-channel protocol provided for exchange traffic (PABX to local exchange signaling) can also be used for tie line traffic between ISDN PABXs (PABX to PABX signaling). This is made possible on the one hand by the fact that both signaling messages and – for connection of PABXs only – signaling sequences have been defined symmetrically by the CCITT, i.e. independently of the direction of signaling between calling/releasing or called/released PABX. In addition, the user-to-user signaling already discussed in Sect. 4.3.5.5 prevents the public ISDN from interpreting a signaling message exchanged via the public ISDN as a message in exchange traffic rather than passing it on to the partner PABX: to this end the PABX-to-PABX ISDN signaling messages for transport via the public ISDN are "packed" e.g. in *USER INFOrmation* messages which serve as transparent transport containers.

Tie lines between ISDN PABXs can either be implemented as private leased circuits bypassing the public ISDN exchanges or routed through the public ISDN exchanges as "nailed connections"; nailed connections can be viewed as semipermanent connections which are established by administrative means in the public ISDN (see OAM in Sect. 6.2.7). From the standpoint of the PABX, nailed connections are effectively permanent once they have been established. The virtual signaling connection assigned to the nailed connection can either be established as a *permanent* signaling connection at the same time as the nailed connection, as is assumed in Fig. 6.18, or can be set up or cleared down by the PABX as required (*temporary* signaling connection).

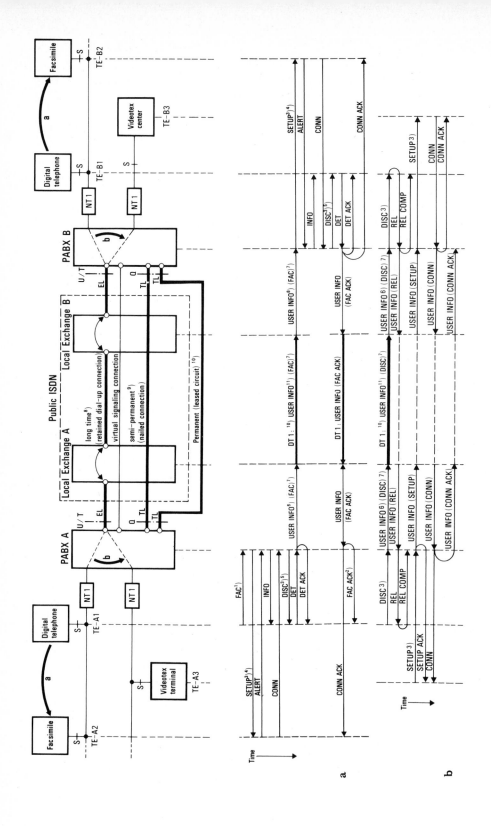

Exchange lines can also be used for tie line traffic between ISDN PABXs to cover peak traffic requirements. In such cases, once an exchange line has been set up using regular call control procedures, it can be used for a *number of* long-distance calls between PABX extensions in succession (Fig. 6.18b), e.g. initially for the connection between TE-A1, 2 and TE-B1, 2 and then for the new connection between TE-A3 and TE-B3. Under such circumstances, the dial-up connection used via the public ISDN is retained for more than one call and is used as a long-time connection between the respective PABXs.

Here the virtual signaling connection assigned to the exchange connection (cf. signaling message *USER INFOrmation*) allows the connection between two PABX extensions (e.g. terminals TE-A1.2 and TE-B1.2) to be cleared down using *DISConnect*, without the exchange connection used in the public network being cleared down at the same time.

Fig. 6.18a, b. Examples of tie traffic between ISDN PABXs. **a** Change of service (in-call modification) for an established 64 kbit/s connection requiring the change of monofunctional terminals, i.e. from digital telephones (TE-A1/TE-B1) to facsimile terminals (TE-A2/TE-B2): the procedure used for extension-to-PABX signaling is in accordance with the respective in-call modification procedure defined for the public ISDN; **b** Use of the same exchange line (retained dial-up connection) successively for the communication links TE-A1, 2/TE-B1, 2 *and* TE-A3/TE-B3.

O——————O	64 kbit/s connection
O————————O	signaling connection
EL	Exchange line
TL	Tie line
NT1	Network termination
TE	Terminal equipment
S, T, U	ISDN reference points (see Sect. 4.1.1)
Q	Reference point applying to PABX-to-PABX connections (with or without intervention of a public ISDN)
SCCP	Signaling Connection Control Part (see Sect. 6.3.4)
ISUP	ISDN User Part (see Sect. 6.3.3)

[1] D-channel protocol (I.451) message FACility requesting change of service for an established call (see Sect. 6.3.6)

[2] I.451 message FACility ACKnowledge indicating successful completion of change of service

[3] I.451 basic call control messages for set-up and clearing of 64 kbit/s ISDN connections (see Sect. 4.3.5 and Fig. 4.23)

[4] Connection establishment to the "new" terminal (TE-A2, Te-B2)

[5] Disconnection (release of B-channel) of the "old" terminal (TE-A1, TE-B1)

[6] I.451 message "USER INFOrmation" enables the transparent exchange of signaling information between users via the public ISDN (see Sect. 4.3.5.5). Under these circumstances it is used as a transport container for private signaling messages ([7]).

[7] Private signaling messages between ISDN PABXs in conjunction with tie line connections via virtual signaling connections

[8] Establishment of the tie line connection via public ISDN exchanges is controlled by the PABX using normal call control procedures, as with regular exchange lines.

[9] Establishment of the tie line connection via public ISDN exchanges is effected by administrative means in the public ISDN.

[10] SCCP message performing end-to-end transport of ISUP signaling messages (see Sect. 6.3.4).

[11] ISUP message "User-to-user information" (see Sect. 6.3.3).

7 Transmission Methods in the ISDN

7.1 Introductory Remarks

According to its CCITT definition [7.1], the ISDN is based on the digitized telephone network. It can therefore use the same digital transmission systems as the telephone network on the interexchange (trunk) circuits. However, in order to use the subscriber lines for ISDN basic accesses at 144 kbit/s (cf. Sect. 4.2.1.2), new transmission methods must be employed, although existing cables can be retained. Moreover, for the B-ISDN access, an optical fiber has to be provided.

The CCITT and CCIR have defined the basic equipment characteristics (interfaces, bit rates, transmission quality) for transmitting digital signals on cables and other media (see Sect. 7.4), and detailed Recommendations exist for multiplexing equipment (Sect. 7.5).

The CCITT has taken account of ISDN requirements since as early as 1976 when considering network synchronization (Sect. 7.6) and transmission quality on digital routes (Sect. 7.7).

7.2 The Hierarchy of Digital Transmission Channels

7.2.1 Basic Building Block: 64 kbit/s

The ISDN is based on the 64 kbit/s channel for the digitized telephone signal.

Because of the importance of the 64 kbit/s channel, let us consider its origin in some detail. It derives from the use of pulse code modulation (PCM) in the telephone network. PCM is an analog/digital (A/D) conversion procedure (Fig. 7.1) in which "samples", i.e. discrete instantaneous values, are taken at a rate of 8000 Hz from the telephone signal, which as in conventional long-distance telephony is limited to a frequency range up to 3400 Hz. The range of signal values to be transmitted is divided into a certain number of "quantizing intervals" (Fig. 7.1 shows only eight intervals for the sake of clarity). The interval into which each sample falls is ascertained and the number of the interval is transmitted in binary form (3-digit code in Fig. 7.1). The more quantizing intervals are provided, the smaller the "quantizing noise" ($QN = S_r - S_0$ in Fig. 7.1).

When PCM was first introduced (around 1962), the number of quantizing intervals was chosen such that the quantizing distortion is virtually inaudible if

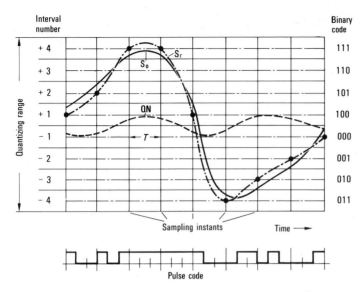

Fig. 7.1. Principle of Pulse Code Modulation (PCM).
S_o original signal, S_r reconstructed signal, QN quantizing noise, T sampling interval (for telephony: $1/8000$ Hz $= 125$ μs). As in CCITT Rec. G.711 the quantizing intervals are numbered from ± 1, while the binary code corresponding to the interval numbers starts from ± 0; the first bit denotes the sign. ● Reconstructed (quantized) sample

in *one* telephone call there are *four* conversions from analog to PCM and vice versa [7.2]. In those days the only conceivable application of PCM was for digital transmission between exchanges using space-division switching. It was found that if a suitable "non-uniform encoding" method [7.3] is used, $128 = 2^7$ quantizing intervals are necessary, i.e. 7 bits must be transmitted for each sample. Later it was realized that in an international connection up to 14 or 15 PCM conversions could be cascaded. The CCITT therefore decided in 1969 to specify 8-bit PCM as the standard; hence the basic unit of 8 bits $\times 8000\, 1/s = 64$ kbit/s.

No single internationally accepted standard has been established for 8-bit encoding; instead, two similar encoding laws have evolved: the "A-law" (used in Europe and most non-European countries) and the "μ-law" (North America, Japan) [7.4]. However, it is possible for systems conforming to different laws to interoperate; in principle, every PCM code word of one law is replaced by that word of the other law which yields the best match of the decoded (i.e. analog) sample.

The distribution of the permissible quantizing distortion between the national portions and the international portion of a telephone connection is covered by CCITT Rec. G.113 [7.5; see also 7.2].

In a telephone connection between two subscribers in the ISDN there is only *one* analog/digital and *one* digital/analog conversion. Although there is therefore no necessity in principle to adhere to 8-bit encoding and hence 64 bit/s, this standard can be usefully applied to the ISDN for the following reasons:

- The multiplexed signals introduced (see Sect. 7.2.2) are based on the 64 kbit/s channel.
- Switching networks in digital exchanges (see Sect. 6.2.3) through-connect 64 kbit/s signals.
- Analog/digital and digital/analog PCM converters – also known as codecs (*co*der + *dec*oder) – for PCM at 64 kbit/s are available in the form of LSI devices.
- In future, improved intelligibility and fidelity of reproduction will be desirable for speech transmission. For this purpose a new telephone service providing a voice frequency bandwidth of about 7 kHz will be introduced in the ISDN (see Sect. 2.3.1.2). A subscriber will be able to use the new service instead of PCM telephony if the other party also has a suitable terminal. A/D conversion of a 7-kHz voice signal can be implemented by means of adaptive differential PCM with a sampling frequency of 16 kHz. In differential PCM, the difference between a "predicted" value (estimated value obtained by extrapolation from preceding signal values) and the actual sample is encoded by means of PCM; the "scale of the quantizing intervals" is thus movable within certain limits. Both the predictor used for generating the estimated value, and the scale of quantizing intervals can be matched *adaptively* to the individual characteristics of the signal present (*adaptive* differential PCM, ADPCM); consequently, the signal can be described with fewer bits than in conventional PCM, e.g. with four bits per sample, giving 16×4 kbit/s, i.e. 64 kbit/s for the 7 kHz voice signal – the same as for the conventional telephone signal using PCM.

   The CCITT has elaborated Recommendation G.722 for 7 kHz voice encoding [7.6].
- Terminals for signals other than voice signals (e.g. for facsimile and data transmission) can also use the relatively high bit rate of 64 kbit/s to advantage.

In the telephone service (300–3400 Hz), conversion from PCM to ADPCM can be employed in the interest of economy, in particular for transmission on satellite or transoceanic cable links. For this purpose, CCITT Rec. G.726 [7.7] defines ADPCM at 32, 24 and 16 kbit/s (and 40 kbit/s for data-modem signals).

   For mobile telephony, even lower bit rates are required since radio bandwidth is at a premium; thus, for instance, in the European digital cellular mobile radio (GSM system) a bit rate of 13 kbit/s will be used (cf. Sect. 8.3).

7.2.2 Primary Multiplex Signals

A de facto standard (from AT & T) for a PCM multiplex system first came into being in the USA, where in 1962 an equipment was put into service which combined 24 digitized telephone signals each encoded using seven bits; to each

7-bit code word was added an eighth bit for signaling, and for every $24 \times (7 + 1)$ bits a further bit for frame alignment was provided, giving a total bit rate of $(24 \times 8 + 1)$ bits $\times$ 8000 1/s = 1544 kbit/s. This frame structure (as shown in Fig. 7.2a) was retained when 8-bit encoding was introduced (see Sect. 7.2.1) [7.8]. Most existing American 1544 kbit/s line systems ("T1 systems") do not allow long sequences of zeros to be transmitted; for the time being, therefore, each octet must contain at least *one* 1-bit. In voice transmission this is easily achieved by avoiding the use of the 8×0 octet, but for non-voice signals it is necessary for the time being to fix the eighth bit (the bit transmitted last) at binary "1". The long-term plan is to use the B8ZS line code (cf. Sect. 7.4.2).

Within CEPT, the Europeans agreed in 1969 on a system for combining 30 encoded telephone signals. Signaling was completely separate from voice transmission from the outset, and ample provision was made for pulse frame alignment, alarm messages, etc. [7.9].

The pulse frame is shown in Fig. 7.2b. Time slots 1 to 15 and 17 to 31 are initially intended for PCM-encoded telephone signals; in the ISDN they can be used for any 64 kbit/s signals originating from the subscriber and transmitted in B-channels (see Sect. 4.2). Time slot 16 was originally intended for channel-associated signaling; in the ISDN's primary rate access channel structure (see

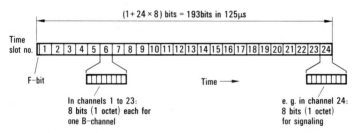

Fig. 7.2a. Pulse Frame of the 1544 kbit/s Signal as Used in the ISDN.
Within a multiframe comprising 24 frames, the F-bit is used for indication of frames and multiframes, for alarm messages and for CRC-6 check bits. These check bits serve to prevent erroneous frame alignment and to measure the bit error ratio

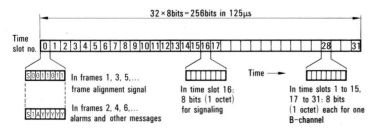

Fig. 7.2b. Pulse Frame of the 2048 kbit/s Multiplex Signal (in the ISDN).
S bit used for multiframing and CRC-4 check, A bit for remote prompt alarm, Y spare

Sect. 4.2.1.2) e.g. between a large private branch exchange and a local exchange in the public network (see Sect. 6.4), it carries a 64 kbit/s D-channel signal (see Sect. 4.2.3), and between public exchanges the signal of CCITT Common Channel Signaling System No. 7 (see Sect. 6.3).

7.2.3 Digital Multiplex Hierarchies

Digital transmission is no different from traditional FDM in that, where large volumes of traffic have to be transmitted, sufficiently large channel blocks must be provided. For this purpose systems of "hierarchical" multiplex levels were developed. These are illustrated in Fig. 7.3. In the system which evolved in Europe, *four* lower-level signals are combined at each stage. The US and Japanese systems use multiplexing factors of 3, 4, 5, and 7. The combining of several "tributaries" to obtain a signal at a higher bit rate is effected by *digital multiplexers*, considered in Sect. 7.5.2.

While having 64 kbit/s as a common base, the hierarchies developed separately up to levels of 139 and 45 Mbit/s. Above these levels, they are now unified to some extent within the *Synchronous Digital Hierarchy* SDH [7.10 to 7.13]. The bit rates of SDH signals will be the same worldwide, and the same multiplexing principles will be used; however, a multiplex signal at, say 622 Mbit/s composed of "American" tributary signals is not compatible with one at the same bit rate comprising "European" signals.

A multiplexer may skip a hierarchical level – e.g. DMX2/34 and M13 in Fig. 7.3; and the function of the SDH multiplexers STM-1 and STM-4 may be combined in one unit.

The fixed points in the hierarchies are the *interfaces*, at which equipments operating at the same bit rate can basically be interconnected as required. The pulse shapes and line codes of the interface signals, etc. are defined in CCITT Rec. G.703 [7.14]. The interface circuits are frequently routed to *digital distribution frames* [7.3] providing flexible interconnection. In future these will largely be replaced by electronic "*cross-connects*" which may also perform multiplexing and protection-switching functions.

Actual transmission via cable or radio links is very often accomplished at the "hierarchical" bit rates. It is also possible first to combine several hierarchical digital signals by additional multiplexing before transmitting them. Examples of this are radio relay transmission at 2×8448 kbit/s, transmission on coaxial lines or optical fibers at 565 Mbit/s, and in North America, transmission over cables at 3152 kbit/s, 90 Mbit/s, or 140 Mbit/s (the latter bit rate being incidentally the same as one of the hierarchical bit rates in Europe). In general there are no internationally standardized equipment interfaces for the resulting non-hierarchical bit rates.

The signals of the SDH are suitable for transmission of broadband signals, e.g. ATM signals (cf. Sect. 4.5).

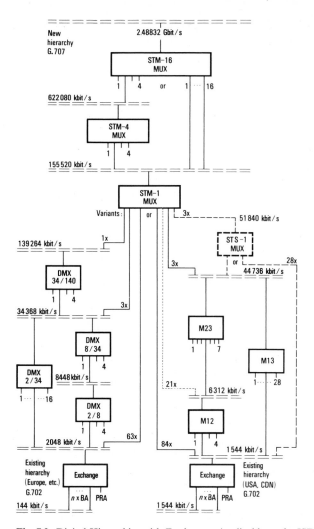

Fig. 7.3. Digital Hierarchies with Equipment Applicable to the ISDN

= = =	Hierarchy level with bit rate	DMX	Digital multiplexer
— — —	Only in North America (not CCITT)	M,MUX	Multiplexer
- - - -	It is unlikely that this theoretical	PRA	Primary rate access
	multiplex relationship will be realized	STM	Synchronous transport module
BA	Basic access	STS	Synchronous transport signal

7.3 Transmission Media

Basically the same transmission media are used in the ISDN as in conventional networks, although conventional copper lines are no longer adequate as subscriber lines for broadband services; optical fibers must be used instead.

7.3.1 Conductors in Cables

Some of the basic characteristics of conductors in cables are briefly described below.

● Symmetrical pairs, i.e. twisted *copper wire pairs* with paper or polyethylene insulation are combined to form cables containing about 20 to 2000 pairs. These are used for transmitting analog voice-frequency signals over subscriber lines and in the local and short-haul network. Since PCM was first introduced, existing cables of this type have also been used for transmitting digital signals. The objective here is to ensure better utilization of cable capacity (for example in the 30-channel PCM system – see Sect. 7.2.2 – instead of 30 wire pairs only two are needed for 30 voice circuits), thus obviating the need to lay new cables to keep pace with increasing traffic volumes.

　　The higher the bit rate of a digital signal, the greater the transmission loss on a wire pair (cf. Fig. 7.8). Crosstalk, i.e. coupling between different wire pairs within a cable, also becomes increasingly troublesome. At bit rates of 2048 or 1544 kbit/s, in order to provide reliable separation of the wanted signal from interference caused e.g. by crosstalk, it is necessary to install *regenerative repeaters* at intervals of approximately 1.7 to 3.5 km [7.15]. For subscriber lines with a net bit rate of 144 kbit/s, regenerators are only required after 8 km if *echo cancelation* is employed (Sect. 7.4.3).

● *Coaxial cables* consist of pairs whose dimensions are characterized by the external diameter of the inner conductor and the internal diameter of the outer conductor. Coaxial pairs with dimensions of 2.6/9.5 mm (*large tube*) and 1.2/4.4 mm (*small tube*) are used extensively for transmitting carrier frequency signals in the long-haul network. They are also suitable for digital transmission. Often spare capacity is available for this purpose in existing cables. Alternatively, cables may be converted from carrier frequency to digital transmission (retaining the same repeater spacing). It is not likely that new coaxial cables of the two types mentioned will in future be laid specially for digital transmission.

　　For economic reasons the 0.7/2.9 mm *mini coaxial tube* has been developed specifically for digital transmission. It is used in some countries for transmitting 2 and 8 Mbit/s signals.

● *Optical fibers*, although a recent development, provide nevertheless the most promising medium for digital transmission and hence for future communication systems in general. Optical fibers will certainly predominate in the cables of future transmission systems – at any rate in newly installed cables; this applies to both long-haul and local traffic.

　　An optical fiber is a silica thread through which light rays are transmitted in the infrared region. Attenuation is a function of the purity of the material. There is however an unavoidable minimum attenuation due to absorption and scattering of the light. Attenuation decreases with increasing wavelength

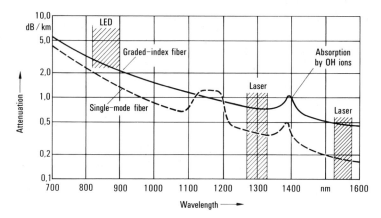

Fig. 7.4. Attenuation (per km) of Typical Optical Fibers as a Function of Wavelength. Hatched areas: wavelength regions of conventional optoelectronic transducers (LEDs and laser diodes)

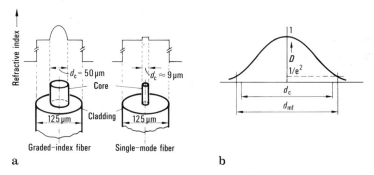

a   b

Fig. 7.5. a The Two Types of Optical Fiber. With the single-mode fiber, the mode field diameter rather than the core diameter is specified – see **b. b** Definition of the Mode Field Diameter d_{mf}. This diameter is about 10% greater than the core diameter d_c. D relative optical power density

(see Fig. 7.4). In practice two types of optical fiber are used (Fig. 7.5):
 – The *graded-index fiber*: due to its relatively large core diameter (50 μm) it is capable of accepting light from light-emitting diodes (LEDs); however, it is essentially limited to transmitting signals up to 140 Mbit/s.
 – The *single-mode fiber*: due to its small core diameter (9 μm) in practice only laser diodes or possibly edge-emitting diodes can be used as light sources.

Details of the technology and application techniques of optical fibers and fiber-optic cables are dealt with more fully in [7.16 and 7.17].

7.3.2 Radio Relay

The transmission medium used by radio relay is free space. According to the laws of radiation, attenuation occurs because there is a limit to the narrowness of any given radio signal beam. The attenuation coefficient (in decibels) does not increase, as in cables, proportionally with the distance but only with its logarithm, so that substantially greater distances can be spanned without amplification than using copper cables. In practice, however, additional attenuation occurs in free space due partly to adsorption in the atmosphere and above all to scattering at raindrops. In practice, rain attenuation only becomes a problem at frequencies above about 10 GHz [7.18]. It then necessitates the use of shorter hop lengths (i.e. shorter distances between the relay stations), and for this reason the higher frequencies are in practice only used for short-haul systems (see Sect. 7.4.4).

On terrestrial radio links, interference due to multipath propagation can be very troublesome. This is caused by reflections from level ground, water surfaces and abrupt transitions between tropospheric layers [7.18].

The transmission capacity of radio relay is limited. In the case of terrestrial radio relay this is because the frequency range is limited due to absorption and because the same frequencies can only be used at different locations if the latter are sufficiently far apart. Within the radio-frequency channel arrangements defined by CCIR, transmission of digital signals is at present realized at bit rates up to 155 Mbit/s, i.e. an SDH bit rate (cf. Sect. 7.2.3), and 2×155 Mbit/s; transmission at 622 Mbit/s may follow.

For radio relay via satellites, the same basic factors apply as to terrestrial radio relay, including attenuation due to rain. In its favour is the fact that radio beams directed at satellites cross the rain layer at a relatively steep angle.

Frequencies up to about 30 GHz are used in satellite systems. However, there is a limit to the number of geostationary satellites which can be positioned in the equatorial plane (at an altitude of some 36 000 km); the present minimum spacing is $3°$, in future it will be $2°$. And so, once again, the total transmission capacity is limited.

A disadvantage of satellite links is the long transmission delay (about 260 ms for a "satellite hop" i.e. ground-satellite-ground). This makes conversation in telephony more difficult (the more so if two satellite links are connected in series) and can provide problems in interactive data communication (see Sects. 3.8.3 and 7.7.3) [7.19].

Radio relay (including satellite systems) enables communication links to be set up quickly; it is particularly useful if cable transmission systems are not yet available or are ruled out for geographical reasons.

7.4 Equipment for Transmitting Digital Signals on Cable and Radio Links

7.4.1 General

For the user, transmission methods and systems should always meet the same quality requirements irrespective of the transmission medium. This applies

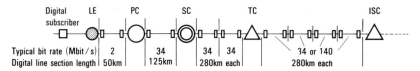

Fig. 7.6. Hypothetical Reference Connection from Subscriber to International Switching Center. LE local exchange, PC primary center, SC secondary center, TC tertiary center, ISC international switching center

especially in the ISDN where every transmission system must be capable of transmitting signals for different services. Network operators, too, are interested in uniform transmission quality, in order to provide flexible interconnection and protection switching of transmission systems. For these reasons, the CCITT has created the notion of the *digital section.*

A digital section is an element in an overall digital link: Fig. 7.6 shows a hypothetical link and its digital sections. According to CCITT definitions [7.3] a digital section includes the whole of the means of digital transmission of a digital signal of specified bit rate between two consecutive digital distribution frames or equivalent (generally using standardized interfaces – see Sect. 7.2.3). The following basic characteristics have been defined for a digital section:

- Length of the hypothetical digital section to which the performance characteristics relate: for the 2 Mbit/s hierarchy, "hypothetical reference digital sections" have been defined with lengths of 50 km (in practice mainly for 2 Mbit/s and 8 Mbit/s) or 280 km (primarily for higher bit rates).
- Bit sequence independence [7.3], i.e. the capability of the transmission system to transmit any bit sequence, including e.g. all-zeros. This requirement is imposed in the ISDN specifically in the interest of unrestricted transmission of text and data signals. Most existing 1544 kbit/s line systems do not meet these requirements at present (see Sect. 7.2.2) but recent systems provide bit sequence independence by using the B8ZS code (cf. Sect. 7.4.2).
- Bit error performance: see Sect. 7.7.1; for example, for a 140 Mbit/s digital section 280 km long, a maximum percentage (0.00045%) of severely errored seconds (i.e. with a bit error ratio exceeding 10^{-3}) is specified.
- Interfaces to other equipment (including the adjacent digital section, if there is any): see Sect. 7.2.2.
- Maximum jitter at input and output. (*Jitter* denotes unintentional but unavoidable phase variations [7.3]; see Sect. 7.7.4).
- Alarm conditions. Alarms must be initiated in the line terminating equipments for multiplex bit rates (from 2 or 1.5 Mbit/s onwards)
  - if the incoming signal is lost: prompt alarm;
  - if the bit error ratio of 10^{-3} in the signal coming from the transmission link is exceeded: prompt alarm (provided that bit error ratio is monitored).

In the case of a prompt alarm, the signal traveling onward (downstream) is replaced by the *Alarm Indication Signal* AIS. With most bit rates, this is an all-ones signal. Only at 44 736 kbit/s does the AIS contain framing bits and certain other overhead bits, and the information bits are replaced by a 1010 . . . sequence. In the USA the AIS was formerly called a "Blue Signal".

For equipment employed in the 2 Mbit/s hierarchy, the internationally agreed characteristics of the digital sections are detailed in CCITT Rec. G.921 [7.20], which applies to both line systems and radio relay systems.

Characteristics of the digital section for the basic access are given in CCITT Rec. G.960 [7.21].

7.4.2 Transmission on Cables in Trunk Circuits

As mentioned in Sect. 7.3.1, transmission on copper wires is mainly realized where cables are already installed and are either used for analog transmission or are still spare.

The future belongs to optical fibers. They have the following advantages over copper:

- Low line attenuation, hence widely spaced repeaters; power-fed intermediate repeaters are seldom used, and in the local network virtually not at all (most intermediate repeaters can be installed in buildings). For high bit rate systems in the long-haul network – including submarine cables – the single-mode fiber is preferred due to its low attenuation (see Fig. 7.4) and large bandwidth.
- Very high transmission capacity (in theory more than 10 Gbit/s).
- Small size, low weight and high flexibility of cable.
- No electrical conductivity, hence no need for protection against electro-magnetic interference, lightning, etc.

Figure 7.7 shows the basic structure of a line system comprising line terminals, cable wires/fibers and repeaters; in fiber-optic systems, repeaters are frequently not required. The arrangement shown in Fig. 7.7 is the physical implementation of a *digital line section*, i.e. a digital section on cable (see Sect. 7.4.1).

Line terminals have the following functions: converting the digital signal from the standardized interface code to the line code (if necessary) and vice versa, monitoring the bit error ratio, alarm indication and power feeding (if necessary).

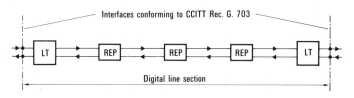

Fig. 7.7. Basic configuration of a Line System for Digital Transmission on a Cable. LT line terminating equipment, REP repeater

Important line codes include:

- *Pseudo-ternary* codes [7.3], i.e. codes in which binary 1 bits are alternately represented by positive and negative pulses, and 0 bits are represented by the voltage 0. The simplest of these codes is the AMI code (alternate mark inversion). It does not provide bit sequence independence. In the HDB3 and B8ZS codes, additional pulses are inserted when a string of 4 or 8 zeros occurs in the original binary signal. This provides sufficient timing even in an all-zeros signal.

  These codes are mainly used for transmission on symmetrical wire pairs.
- *Codes with reduced line digit rate*: 4B/3T (4 bits replaced by 3 ternary signal elements) and 2B/1Q (2 bits replaced by one quaternary signal element). 4B/3T is used on coaxial lines and in some systems for subscriber lines, and 2B/1Q on subscriber lines (cf. next section).
- 5B/6B and 7B/8B: These are redundant binary codes for optical fibres; to every five or seven information bits a sixth or eighth bit is added for monitoring and to provide bit sequence independence.

7.4.3 Transmission on Subscriber Lines

General

The copper wire pairs of the subscriber lines represent a substantial proportion of the overall capital expenditure on the telephone network. It is intended to continue using them in the ISDN so that no new capital investment in subscriber lines is necessary. As with the balanced pairs of the paper- or polyethylene-insulated trunk circuits, it must be remembered that subscriber lines were originally intended for transmitting voice frequency signals (cf. Sect. 7.3.1).

Figure 7.8 shows the attenuation-frequency response of typical wire pairs (with 0.4 and 0.6 mm diameter). For instance, the Deutsche Bundespost Telekom uses 0.4 mm wires for subscriber lines up to 4.2 km long; where greater lengths are involved, practically the same attenuation can be achieved in the voice frequency range for line lengths between 4.2 and 8 km by the appropriate use of 0.6 mm wires for part or all of the line. However, the attenuation increases somewhat at higher frequencies (e.g. by about 3% at 60 kHz between 4.2 and 8 km).

Other administrations or operating companies have similar planning guidelines for subscriber lines.

Figure 7.9 shows the frequency distributions of subscriber line lengths in the Deutsche Bundespost Telekom area. This distribution can be considered as typical for many countries. It shows that 99% of the subscriber lines are no more than 8 km long, i.e. they can be implemented using 0.4 and 0.6 mm wires with no additional measures being necessary. For digital transmission on subscriber lines the range should be just as long. For lengths greater than 8 km a higher supply voltage is required in the case of voice frequency use; for digital

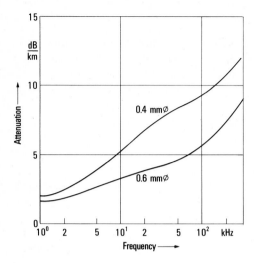

Fig. 7.8. Attenuation Response of Typical Wire Pairs in Local Cables with Polyethylene Insulation

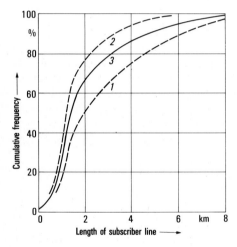

Fig. 7.9. Typical Frequency Distribution of Subscriber Line Lengths (refers to Deutsche Bundespost Telekom Area).

1 Small local networks
(< 800 main stations)
2 Large local networks
(> 10 000 main stations)
3 Average of all local networks

transmission a regenerative repeater may be necessary, or remote multiplexers can be used (cf. Sect. 7.5.2).

In some countries, very long voice-frequency subscriber lines are fitted with loading coils. Loaded lines have very high loss above the VF range; consequently the loading coils must be removed for digitization.

Where there are no existing subscriber lines, laying optical fiber cables can be considered. These are also suitable for "broadband" access (above 2048 kbits/s). In certain rural areas radio relay links may be an option. However, only transmission over copper wire pairs is considered in detail below.

Transmission Method for the Basic Access

In addition to Rec. G.960 (cf. Sect. 7.4.1) for the digital section (i.e. a "black box"), the CCITT has established G.961 [7.22] which defines a number of requirements for subscriber line systems (including maintenance functions). Line codes are specified only in non-normative Appendices to Rec. G.961.

For the ISDN basic access (see Sect. 4.2.2) the information of two B-channels and one D-channel is to be transmitted, together with signals for frame alignment and possibly multiframe alignment as well as maintenance informa-tion, giving an actual total bit rate of about $(2 \times 64 + 2 \times 16)$ kbits/s $=$ 160 kbits/s.

As only one wire pair is available to a subscriber, the digital signals – just like the voice frequency signals in conventional telephony – must be transmitted over this pair in both directions (see Fig. 7.10).

The problem here is that the receiver picks up not only the wanted signal from the distant end, but also, due to unavoidable reflections e.g. at the hybrid junctions a and b (Fig. 7.10), the signal which has been transmitted at *the same* end but which causes interference in the receiver. This reflected signal (somewhat imprecisely termed "echo") must be neutralized. In practice there are two possible options:

- The first method studied and implemented was the *time division method* (also known as the burst or pingpong method) whereby information blocks are formed, each containing e.g. two octets from the two B-channels and four bits from the D-channel; these blocks are transmitted *alternately* in both direc-tions as shown in Fig. 7.11. Between the end of transmission of a data block, e.g. at location A, and the start of transmission at the other end of the line (e.g. at location B) there must be an interval T_i in which disturbing reflections can decay. The length of that interval must be somewhat greater than the signal delay T_s. The greater length of the subscriber line and hence of the signal delay T_s, the less time is available for transmitting the data block, and hence the higher the transmission rate required during actual transmission. Fig. 7.12 shows transmission rate versus line length in the case of four B-octets, etc. being transmitted per block (period $T = 250$ µs as in Fig. 7.11). It shows that a transmission rate of 550 kbits/s would be required for a range of 8 km. Using this high rate poses problems due to the high line attenuation,

Fig. 7.10. Two-Wire Duplex Transmission: the signals for both transmission directions are carried on the same wire pair.
T transmitter, R receiver, 1) interfering reflections of the signal transmitted in A

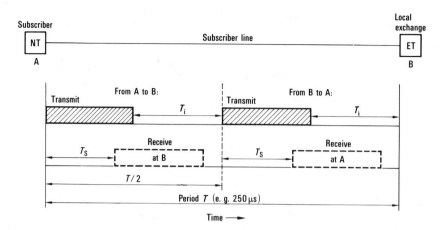

Fig.7.11. Principle of the Time Division Method.
ET exchange termination, NT network termination, T_i interval, T_s signal delay, LE local exchange

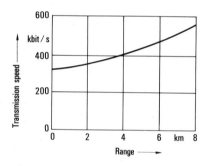

Fig. 7.12. Transmission Rate Required for the Time Division Method as a Function of the Range (applies to $T = 250 \,\mu s$)

crosstalk between adjacent wire pairs and radio interference. The method can, however, be used for short subscriber lines, e.g. in private branch exchange networks.

● *Echo cancelation* avoids the above-mentioned problems associated with the time division method. It is the standard in many networks including those of the USA and Germany. With this method the transmission rate is independent of the range: transmission takes place continuously. Reflections at hybrid junctions a and b (Fig. 7.10) or at discontinuities on the route cause interference directly. In order to neutralize the reflected signal it must be very precisely simulated and subtracted from the received signal, so that only the wanted signal is left. For example, if the line attenuation is 35 dB (this corresponds to 4.2 km with 0.4 mm wire diameter at 60 kHz, see below), the residual reflected "own" signal must be attenuated by about 55 dB by canceling. The reflected signal must therefore be precisely simulated to within approximately 2%. This is possible using an *echo canceler* providing automatic, adaptive adjustment to the characteristics of the line [7.23]. A canceler of this type can only be implemented cost-effectively as a digital LSI circuit.

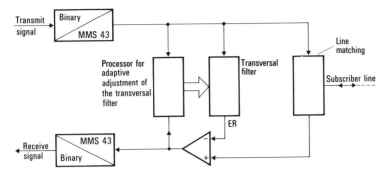

Fig. 7.13. Basic Configuration for Transmitting a 160 kbit/s Signal on Subscriber Lines (echo canceling method). ER output signal from transversal filter (representing echo replica), MMS 43 line code (see Sect. 7.4.2)

Figure 7.13 shows the basic setup of a transmission equipment incorporating echo cancelation. The transmitter feeds the signal to the transmission line on the one hand, and to the echo canceler on the other. The latter is designed as a transversal filter [7.24] whose coefficients adjust adaptively such that a replica ER of the echo signal is produced at the output of this filter. This replica is subtracted from the received signal. This function is similar in principle to that of the echo cancelers used in the analog telephone network to neutralize the speaker echo in intercontinental traffic or on satellite links [7.25].

Implementation of the digital echo canceler is facilitated by selecting a suitable line code. Deutsche Bundespost Telekom has chosen the MMS 43 code, a particular 4B/3T code (cf. Sect. 7.4.2). It provides a relatively small low-frequency portion of the line signal; in addition, the line digit rate is reduced to 120 kbauds. The signal power is then concentrated around 60 kHz. The line attenuation is thus reduced and the crosstalk attenuation increased.

In the USA the 2B/1Q code has been standardized [7.26]. This reduces the line digit rate to 1/2 of the effective bit rate.

As long as copper wires are used for the subscriber lines, it is advisable for the network termination NT1 (or possible NT12, cf. Sect. 4.2.2.6) and perhaps also the terminal equipment (especially a telephone) to be powered from the exchange at least for emergency operation. Power-feeding of the NT1 is mainly advantageous for fault diagnosis using test loops under control of the exchange.

In the idle condition, i.e. while no communication is taking place, the network termination is not fully active, and so the full supply current is not continuously required. However, the network termination NT1 is permanently ready to receive an "activation signal" – a specifically defined pulse train (cf. Sect. 4.2.2.5). Once it is received, NT1 is activated, i.e. all the functional components are switched on

Multiplexed Signals

Subscribers with *multiple channel access* – mainly large digital private branch exchanges – are normally connected to the exchange via 1544 or 2048 kbits/s transmission systems (Table 7.2). The associated pulse frames have already been described in Figs. 7.2a and 7.2b in Sect. 7.2.2. Time slot 24 (or 16) of the multiplexed signal contains a 64 kbits/s D-channel signal.

The transmission method is then essentially the same as on the inter-exchange trunks (Sect. 7.4.2); regenerators must be used on the copper wire pairs of existing subscriber lines if, for example, in the case of 0.4 mm wires the line length exceeds about 1.5 to 1.9 km, and with 0.6 mm wires, 2.1 to 2.7 km.

In many cases it is advisable to combine the signals of remote subscribers (especially those whose line to the local exchange would be longer than 8 km) using a remote multiplexer or concentrator. For instance, the signals from 12 basic accesses can be combined to form a 2048 kbits/s signal, the associated D-channel signals being accommodated separately in specific time slots (which requires 192 kbits/s).

7.4.4 Radio Relay Transmission

Whereas analog radio relay systems have been used mainly in the toll network, digital radio systems are being deployed increasingly in the *short-haul network* and in the *local network*. The systems used in the short-haul and local network mainly employ radio frequency ranges around 13 and 15 GHz, in the local network also around 18 GHz [7.27], and more recently also about 40 and even 55 GHz. Between 13 and 40 GHz, the hop length decreases from about 25 to 5 km, depending on circumstances such as rain fall rate.

As in cable systems, higher bit rates (namely 34, 45, 90, 140 and 155 Mbit/s – cf. Sects. 7.2.3 and 7.3.2) are used for transmission in the *toll network* due to the larger channel blocks. The existing infrastructure of the analog radio relay network, i.e. primarily the existing system of radio relay towers located between 30 and 70 km apart, can be used by digital systems. In practice, systems in 1900, 3900, 4700 and 6700 MHz bands as well as around 11 GHz are in operation.

7.5 Multiplexed Signals and Multiplexing Equipment

7.5.1 Synchronous Multiplexed Signals

The 2048 and 1544 kbits/s signals have already been described in Sect. 7.2.2; their pulse frames are shown in Fig. 7.2. For practical reasons (use of identical components for transmitters, receivers and monitoring equipment with all signal types), CCITT Rec. G.704 [7.28] specifies the following:

- All 2048 kbits/s signals must have the same frame length (256 bits) and the same bits 1 to 8 in time slot 0 of the frame (see Fig. 7.2b)[1] (the Y bits excepted). This gives a frame repetition frequency of 8 kHz.
- All 1544 kbits/s signals must have a frame length of 13 bits and the same F bit. The frame repetition frequency is likewise 8 kHz.
- All 2048 or 1544 k bit/s signals with octet structure (i.e. signals containing consecutive groups of eight bits) must have a pulse frame with the octet arrangement shown in Figs. 7.2b or 7.2a.
- Similar rules apply to 8448 and 6312 kbits/s octet-structured signals.

Octet structured signal sources in the ISDN include multiplexers for subscriber signals, concentrators and digital exchanges.

7.5.2 Digital Multiplexers

Digital multiplexing equipment is used to form multiplexed signals with bit rates above 2048 kbit/s.

Multiplexers with output bit rates up to 139 Mbit/s (see Fig. 7.3) are specified in CCITT Recs. G.742 and G.751 [7.29] for the 2 Mbit/s hierarchy, and in G.743 and G.752 [7.30] for the 1.5 Mbit/s hierarchy. These are so designed that they can combine signals of different origin, i.e. signals with, for instance, different frame structures; these signals may be *plesiochronous*, i.e. with bit rates which are nominally identical but may actually deviate from the nominal value within a certain tolerance range (e.g. $\pm 5 \times 10^{-5}$ at 2048 and 1544 kbit/s).

The bit rate of the output signal of a multiplexer of this kind is determined by an autonomous crystal-controlled generator.

Positive justification ("positive stuffing") is employed for the multiplexing process: each input signal is assigned a transmission capacity *greater* (e.g. by 0.2 percent) than its nominal bit rate. The input signal is written into a buffer store with its own timing signal and read out again by a higher-frequency timing signal corresponding to the transmission capacity. As soon as the phase difference between the write timing signal and the read timing signal exceeds a certain amount, a justifying digit containing no information is inserted in the output signal, thus providing a degree of "breathing space".

Figure 7.14 illustrates this principle. In the (simplified) example, if the phase difference between input and output signal has increased by *one* bit interval, a justifying digit is inserted. In practice, justifying digits are inserted only at

---

[1] There may be exceptions for "point-to-point" connections. However, these are irrelevant to the ISDN.

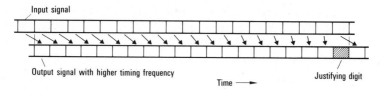

Input signal

Output signal with higher timing frequency

Time ⟶

Justifying digit

Fig. 7.14. Principle of Positive Justification, as Used in Digital Multiplex Equipment for Timing Alignment

specific positions in the pulse frame. The demultiplexer (receiver) is notified whether or not justification has taken place by means of "justification service digits".

The demultiplexer uses the justification service digits to reconstruct each input signal with no loss of information, i.e. with the original bit rate.

Multilation of the justification service digits causes inadvertent omission or repetition of a bit, i.e. a *slip* (cf. Sects. 7.6.1 and 7.7.2). To avoid this, three or five identical justification bits, distributed over the pulse frame, are transmitted, and "majority decision" is used at the receiver.

In the new Synchronous Digital Hierarchy SDH, *positive/zero/negative justification* [7.3] is used for most multiplex relations. The input signal is assigned a transmission capacity nominally *equal* to the bit rate of this signal. If the actual bit rate of the input signal is smaller than the transmission capacity provided, positive justification is performed. If the input bit rate exceeds the regular transmission capacity, then negative justification is employed: The "excess" information is accommodated in an auxiliary channel. The procedure is more complicated than positive justification but makes synchronous operation possible.

7.6 Network Synchronization

7.6.1 Necessity for Network Synchronization

Two or more digital equipments are said to operate *synchronously* if they have the same timing frequency and hence a fixed phase relationship to each other. The main consideration here is to synchronize the bit timing; once this is achieved, frame alignment is easily established using the *frame alignment signals* (Fig. 7.2). In practice, ideal synchronization is impossible to achieve, especially if equipments are far apart geographically. It is sufficient for the equipments to operate *mesochronously*, i.e. with the same average frequency over time; the phase difference between their respective timing signals can accordingly fluctuate (normally within specified limits). It is customary to describe a network as operating synchronously even if strictly speaking its operation is mesochronous.

If the transmitter and receiver of a digital signal are not synchronous, disturbances occur: the incoming signal is fed at its own timing frequency f_1 into a buffer store and read out at the different local timing frequency f_2. If this is higher than the frequency of the incoming signal, the store is emptied "too quickly", with the result that one or more bits are read out again (i.e. repeated) as soon as the phase difference between the two timing signals has reached a magnitude corresponding to the store capacity.

If the local timing frequency is lower than that of the incoming signal, the buffer memory is emptied "not quickly enough"; as soon as the phase difference reaches the critical magnitude, one or more bits of the incoming signal are skipped, i.e. they are lost.

In both cases the event is known as a *slip*. Slips occur mainly at the inputs to digital exchanges. The multiplex signal inputs of digital exchanges for 1.5 and 8 Mbit/s are equipped accordingly: their buffer stores are so designed that, if a slip occurs, one pulse frame (for 1.5 and 2 Mbit/s: see Fig. 7.2) of the multiplex signal is repeated or is lost, i.e. one octet (= 8 bits) for each 64 kbit/s signal. In PCM systems (see Sect. 7.2.1) this corresponds to a PCM code word.

The pulse frame or PCM code word occurs every 125 μs. Accordingly a slip occurs as soon as the time instant marking the start of the pulse frame of the incoming signal (with timing frequency f_1) has shifted by 125 μs relative to the corresponding time instant of the local timing system operating at f_2. The mean interval between two slips is therefore

$$T_s = \left| \frac{125 \; \mu s}{(f_1 - f_2)/f_2} \right|. \tag{7.1}$$

As the slips produce disturbances of varying magnitude (discussed in Sect. 7.7.2), it is desirable to eliminate them as far as possible. All the clocks in a digital network are therefore at least nominally operated synchronously. Slips cannot be totally prevented in this way, because from time to time a digital exchange may be unable to run synchronously (Sect. 7.6.3), and in digital multiplexers slips can occur as a consequence of bit errors (Sect. 7.5.2). Slips will also occur in traffic between two networks each of which is synchronized independently (e.g. two national networks).

7.6.2 Achieving Network Synchronization

A synchronized network will always be limited geographically, usually to the size of a country such as Germany. For practical reasons, a country may also be divided into several regions, each independently synchronized, or several telecom carriers may each have a separate synchronous network. Alternatively, two or more small countries may implement a jointly synchronized network.

In the past, several methods of network synchronization have been discussed and some have been analyzed in great detail; however, in the following we shall only describe the method which in practice is likely to be used virtually exclusively, namely the *master-slave method*. In this method a primary reference clock (the *master*) controls all the exchanges directly or via intermediate stages and thus determines the frequency of all the 64, 1544 and 2048 kbit/s signals in the network. In the future, with the evolving broadband ISDN and in conjunction with new synchronous multiplex structures, synchronization will be extended to higher bit rates.

The reference clock is a cesium-beam oscillator with a frequency uncertainty of not more than $\pm 10^{-11}$ (cf. Sect. 7.6.3). It is preferably located near the geographical center of the synchronized network, e.g. in Germany at Darmstadt, and for the network of the AT & T at Hillsboro (Missouri). The reference clock passes on its frequency to other equipments either as 2048 kbit/s or 1544 kbit/s signal timing or, where digital links are not yet established, using FDM *frequency comparison pilots*. (The Hillsboro clock delivers a 2048 kHz analog signal). Suitable reference timing can also be derived from radio signals of the LORAN-C or OMEGA navigation systems.

The synchronizing signals are generally distributed from the top down in accordance with the switching network hierarchy, as shown in Fig. 7.15 (cf. Sect. 3.2.1 and Figs. 3.2, 3.3). This should not be regarded as a rigid principle; the illustration also shows an example of a possible direct path bypassing one hierarchy level (a).

If a primary synchronization path normally used for top-down synchronization fails, if possible a secondary path should be available. It may originate from the same network node or exchange as the normal synchronization path but via a different transmission route (b), or else from a different exchange (c).

A local exchange connected only to *one* primary center does not need a secondary path for synchronization, because if the digital signal from the primary center fails, it can anyway handle only local digital traffic, for

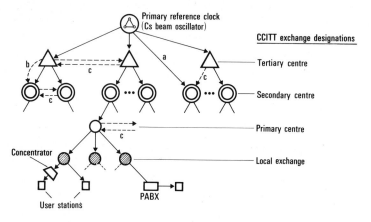

Fig. 7.15. Hierarchical Structure of a Network Synchronization System.

⟶   Normal synchronization path
--->   Secondary synchronization path
a   direct path
b, c   see text
PABX   Private automatic branch exchange

which the frequency accuracy of the free-running local exchange (tolerance typically $\pm 10^{-7}$) if adequate.

Private branch exchanges as well as digital subscriber stations are synchronized from the public network, i.e. generally from a local exchange.

7.6.3 Clock Supply Requirements

CCITT Rec. G.811 [7.31] gives guidelines for the synchronization of national networks in respect of ISDN requirements, particularly as regards international interworking. The recommendation provides for primary reference clocks as mentioned above with a frequency deviation of up to $\pm 10^{-11}$.

In practice, a generator of this accuracy can only be implemented as a caesium frequency standard. As the clock of every synchronized digital network – for example a national network – is determined by a cesium standard of this kind, the average interval between two slips in international traffic is theoretically not less than $125 \, \mu s/2 \times 10^{-11}$ = approximately 70 days (from Eq. 7.1).

CCITT Rec. G.823 [7.32] specifies a maximum value for the phase variation at the output of a network node (i.e. generally an exchange) handling international traffic (10 μs). This variation is with respect to a hypothetical signal with the mean frequency of the caesium reference clock. In traffic between two networks, the phase variations are superimposed on the regular drift of the two reference clocks; in addition, signal delay variations on the intervening circuit can occur. The consequence of both effects is that the actual interval between two slips may be greater or less than 70 days.

While CCITT Rec. G.811 while specifies the performance of the primary reference clock, Rec. G.812 [7.33] defines the behaviour of synchronized clocks ("slave clocks") in network nodes – i.e. essentially clocks in digital exchanges. In

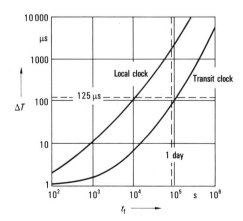

Fig. 7.16. Permissible Relative Time Interval Error of a Slave Clock in Accordance with Rec. 6.812.

particular, the "holdover mode" (when the synchronizing timing input has failed) is specified. The characteristic magnitude is the "time interval error", i.e. the phase deviation of the slave clock with respect to the reference timing. Fig. 7.16 shows the allowable deviation, ΔT, as a function of the time, t_f, that has elapsed after the start of the free-running condition.

For economical reasons, a local clock is allowed to have a larger phase deviation than a transit clock. In any case, in the event of failure of the synchronizing timing, a local node will in general only be able to handle local traffic (cf. Sect. 7.6.2).

7.7 Disturbances and Transmission Performance

The transmission performance on a digital link is characterized by four variables: bit errors, slips, transmission delay and phase jitter. Their undesirable effects are considered below with particular reference to the ISDN.

7.7.1 Effect of Bit Errors

Just as in the analog telephone network noise interference is unavoidable but can be tolerated if it stays within certain limits, so in the digital network it is inevitable that bits in the digital signal will be multilated by interference (1 becomes 0 or vice versa). This can also be tolerated if it does not occur too frequently.

Bit errors are either caused by external interference or by thermal noise. *External interference* includes:

- Dial pulses or other pulses on wire pairs in the same cable which are not yet operated digitally: these pulses affect the disturbed wire pair through unavoidable electromagnetic coupling.
- Crosstalk between wire pairs transporting similar signals. (In principle this effect can be prevented with careful planning.)
- External electromagnetic effects due e.g. to railways with electrical traction, especially if thyristor control is employed.

These disturbances primarily affect symmetrical pairs in the local network and on subscriber lines. These are therefore assigned a comparatively large proportion of the total bit errors permissible in a connection between two subscribers (see the discussion of CCITT Rec. G.821 below). Bit errors very often occurs in bursts; depending on the circumstances, an error burst can affect between 2 and 50 or in some cases even more consecutive bits.

In coaxial cables, optical fibers and radio relay links, *thermal noise* is a major cause of bit errors. These have a purely random distribution (Poisson distribution) and are amenable to precise planning, which in practice is a compromise between optimum transmission quality and cost considerations.

Bit errors affect individual services in different ways:

- PCM speech transmission: a bit error ratio or error-burst rate of 10^{-5} can be tolerated. Even random (Poisson) errors with a frequency of 10^{-4}, such as may occur briefly in telephony via satellites, only cause slight clicks (not continuous noise).
- Data transmission with error detection (methods using Automatic Repeat Request ARQ such as High Level Data Link Control HDLC [7.34]): if a single bit error or an error burst occurs in a data block, this is detected by the data receiver, which then causes the data block to be repeated. In order to ensure that the effective data throughput is not excessively reduced due to repetitions (i.e. reduced by no more than 10 to 20%) – and this also applies to long data blocks or a long transmission delay (via satellite) [7.19] – a bit error ratio of less than 10^{-6} is desirable.
- Text transmission: modern text transmission, e.g. teletex, employs protected data blocks; the same comments apply as for data transmission.
- Facsimile transmission: the system mainly considered for ISDN use is ISDN telefax (Sect. 2.3.1.2); here, too, block by block protection is provided by the HDLC method, so that as far as the effect of bit errors and the resulting requirements are concerned, the same comments apply as to data transmission (also applicable to ISDN textfax).
- Common channel signaling according to CCITT Signaling System No. 7 (cf. Sect. 6.3): this is also a special case of block by block data transmission using HDLC. Owing to this data protection, the probability of an incorrectly established call is several orders of magnitude less than the frequency of a block disturbance.

CCITT Rec. G.821 [7.35] takes into account the described requirements for services. It also specifies the bit error performance with which future new services of the 64 kbit/s ISDN will have to operate. For an international 64 kbit/s reference connection between two subscribers the recommendation specifies:

- *Severely errored seconds*: fewer than 0.2% of one-second intervals should have a bit error ratio worse than 10^{-3}.
- *Errored seconds*: fewer than 8% of one-second intervals should have any errors.

The "degraded minutes" specification in the extant G.821 has recently be deleted. For the errored seconds and one half of the severely seconds objectives, Rec. G.821 gives an allocation to three "circuit classifications" as shown in Fig. 7.17. The classification contains some degrees of freedom to allow for different conditions in countries of different sizes. For "medium-sized" countries such as Germany, it is advisable to allocate the subscriber line the 15% quoted as an example in the figure, and to extend the "medium grade" quality as far as the primary center (class 4 office) or secondary center (class 3 office).

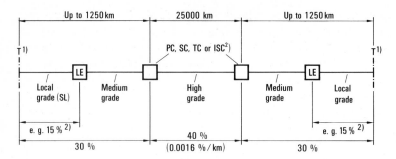

Fig. 7.17. Distribution of Errored Seconds and Severely Errored Seconds Among the Portions of a Complete 64 kbit/s Connection.

SL   Subscriber line   SC   Secondary center
LE   Local exchange   TC   Tertiary center
PC   Primary center   ISC   International switching center
[1] ISDN reference point T (see Sect. 4.1)
[2] depends on choice by administration or carrier

Other relevant details of CCITT Rec. G.821 include the following:

● A share of 20% of the (severely) errored seconds is assigned to a satellite section (within the "high grade" portion).

These requirements apply initially to a 64 kbit/s channel. The requirements concerning severely errored seconds are also directly applicable to higher bit rates. Seconds in which the receiver of the signal under consideration (e.g. a 2048 kbit/s signal) has lost frame alignment are counted as severely errored seconds.

Additional specifications (but consistent with Rec. G.821) for digital sections (cf. Sect. 7.4.1) in the medium and high grade portions of Fig. 7.17 are given in Rec. G.921 [7.20], and specifications for high-speed connections (> 2 Mbit/s) will be given in new Rec. G.82X (under study).

7.7.2 Effect of Slips

Slips (see Sects. 7.5.2 and 7.6.1) should be regarded in the same way as other unavoidable disturbances. Every service can be affected by them, and their effect depends on the nature of the signal:

● PCM speech transmission: a slip produces a phase and amplitude shift, which is usually inaudible or at most can be heard as a click. In practice about 20 slips per minute are permissible.
● Data transmission with error detection (ARQ method): just like a bit error, a slip causes a data block to be received incorrectly. The data block must then be repeated. Slips may therefore occur no more frequently than bit errors.

Because the occurrence of slips can be kept well under control, it appears reasonable to require that slips occur at no more than one tenth the frequency of bit errors. As CCITT Rec. G.821 (see Sect. 7.7.1) specifies a maximum of four bit errors in 90% of all 1-minute intervals, about four slips in ten minutes would be permissible.

- Text and facsimile transmission: the remarks on the effect of bit errors (Sect. 7.7.1) similarly apply here: the undesirable effect is in principle the same as for data transmission and gives rise to the same requirements.
- Common channel signaling complying with CCITT Signaling System No. 7 (cf. Sect. 6.3): here, too, slips have the same effect as bit errors. It is only if the data block disturbance is not detected by the receiver – highly unlikely due to the error correction procedure (see Sect. 7.7.1) – that e.g. an unwanted connection could be set up.
- Transmission of data in the form of multiplexed signals (signals conforming to CCITT Recs. X.50 and X.51 [7.36] or X.22 [7.37] comprising e.g. 20 tributary signals each of 2.4 kbit/s): in the ISDN, signals conforming to CCITT Recs. X.50 and X.51 may occur only on leased lines used specially for data communication. Signals conforming to CCITT Rec. X.22 on the other hand are transferred from subscriber to subscriber. With multiplexed signals, a slip causes the receiver frame alignment to be lost. The receiver must therefore recover frame alignment. All the individual signals within the multiplexed signal are affected and it is even possible that an entire data switching center may be put on alarm status. Slips should therefore only occur at intervals of at least 15 minutes.

CCITT Rec. G.822 [7.38] takes into account both the known requirements of the services and also what is possible in practice. This Recommendation, like CCITT Rec. G.821 (Sect. 7.7.1) applies to a 64 kbit/s connection in the ISDN, which can be 27 500 km long. For such a connection, Rec. G.822 specifies that in at least 98.9% of the time the average interval between two slips shall be not less than 4.8 hours and for virtually the rest of the time (about 1% max.) this interval should exceed 2 minutes. (The slip rate may be even higher for up to 0.1% of the time.) The timing accuracy of digital exchanges must be set accordingly, even under free-running conditions (when clock control "from above" had failed due to a fault) (see Sect. 7.6.3).

7.7.3 Effect of Signal Delay

Signal delay does not generally receive as much attention as bit errors and slips. However, it is also an important factor, especially in the following cases:

- In *voice communication*, as the signal delay increases, conversation becomes more difficult. The CCITT has therefore laid down in Rec. G.114 [7.39] a maximum signal delay for telephony of 400 ms (for *one* transmission direction); this means that *one* satellite section (signal delay T_D: about 260 ms) is permitted. The 400 ms should only be exceeded in exceptional circumstances, e.g. if no connection at all can otherwise be set up.

- In *data transmission* with ARQ (see Sect. 7.7.1), buffer stores must be available at least in the data transmitter, the store size K corresponding to the number of bits that can be transmitted within $2T_D$, i.e. twice the signal delay time: $K = 64\,\text{kbit/s} \times 2T_D$. In the case of data transmission with block-by-block acknowledgement, a long signal delay (e.g. via satellites) can considerably prolong the actual duration of an individual connection.

7.7.4 Effect of Jitter and Wander

As already described in Sect. 7.6, absolutely rigid synchronization of all signals in the digital network is impossible to achieve in practice; frequency variations within a certain range are unavoidable. A frequency variation (i.e. unintentional frequency modulation) can always be described as a phase variation (phase modulation).

Relatively rapid phase variations (with a frequency greater than 20 Hz) are known as *jitter*, slower variations as *wander*. Jitter is mainly attributable to imperfect timing recovery in regenerators of line systems (Sect. 7.4.2), wander to justification processes in digital multiplexers (Sect. 7.5), control errors of phase-locked loops and temperature-dependent signal delay fluctuations in cables.

As jitter and wander are to a certain extent unavoidable, tolerance diagrams have been established which specify, for all the interfaces in the digital hierarchy (see Sect. 7.2 and Fig. 7.3) and for the S/T interface (Sect. 4.2.2), how much jitter or wander is permissible at the input of a unit without causing bit errors, slips or other disturbances. To check the actual jitter/wander tolerance of an input, a pseudorandom test signal with sinusoidal phase modulation is used; a suitable test signal generator (and a jitter measuring set) is specified in CCITT Rec. O.171 [7.40]. It uses e.g. at 2048 kbit/s a signal with a period of $2^{15} - 1$ bits. If a unit can tolerate the test jitter at the input, then it can be assumed that the jitter present during actual operation (and which naturally has no sinusoidal modulation) will not produce disturbances.

Jitter and wander as defined in the interface specifications [7.32] can always be controlled and hence do not impair the onward signal.

7.7.5 Performance of 64 kbit/s Connection Types in the ISDN

CCITT Rec. I.340 [7.41] specifies *connection types* used to implement network connections for the individual services. Each connection type is described by a number of attributes including transmission performance parameters (cf. Table 6.1).

For the *transparent* connection type, the bit error performance will comply with CCITT Rec. G.821, and the slip frequency with CCITT Rec. G.822; bit integrity is also guaranteed, i.e. no bit or octet in the 64 kbit/s signal is intentionally changed e.g. by recoding from 64 to 32 kbit/s (see below).

Non-transparent connections will probably only exist for telephony. Bit manipulations are possible here, in particular:

- Conversion from the PCM code in accordance with the A-law (Sect. 7.2.1) as used in Europe to the PCM μ-law code used e.g. in the USA, or vice versa; this means replacing one octet (PCM code word) by another.
- Conversion from PCM to ADPCM to reduce the bit rate to 32 kbit/s (see Sect. 7.2.1).

8 ISDN – The User's View

The evaluation of the ISDN from the user's point of view is intended to show the ways in which the ISDN improves communication in terms of meeting user needs and supporting new applications. It also considers those requirements that will for some time have to remain unsatisfied. In general, it should be emphasized that the ISDN improves communication through simpler operation, improved accessibility as well as easier and faster access to information. These improvements ultimately result in a better communications cost/ performance ratio, a speeding-up of communication processes (and therefore greater productivity) and an improvement in the quality of decisions through being better informed. The three main groups of users, each with their own profile of communication requirements, will be considered separately: people in the office, people at home and people on the move.

8.1 ISDN in the Office

8.1.1 Telephone Communication

In spite of increasing information exchange via text and data systems, the telephone will continue to be the most frequent application of communications technology in the office sector for some time to come, with a considerable proportion of the working day being spent making telephone calls (approx. 15% [8.1]). Consequently, improvements in this area are of particular importance.

Conventional private automatic branch exchanges (PABX) already provide efficient, high-convenience telephony [8.2]. Here, the new features offered by the ISDN (see Chap. 2) will improve telephone communication still further [8.2]. It is important to remember that with the implementation of the public ISDN the use of these sophisticated features will not be limited to the use within private branch exchanges, as a considerable proportion of business voice traffic goes beyond the confines of private branch exchanges.

The improvements brought by the ISDN to telephone communication relate particularly to call establishment and the accessibility of call partners. For example, the features *on-hook dialing* and *repeat dialing* simplify the call set-up procedure. It is also possible to store the caller's number transmitted by the ISDN in the voice terminal of the called subscriber. This is extremely useful if the user has to return the call: he does not need to make a note of the number, but can activate it at a press of a button.

However, the service attributes that bring about greater accessibility are likely to be of even greater importance for the user. In business traffic especially it is all too often the case that either no-one answers or someone other than the intended person answers, or that the line is busy. A number of studies have been conducted on the subject of the accessibility of called parties (e.g. [8.1, 8.3, 8.4]). These showed that in the business sector only 30% of all call attempts are successful in the sense that the wanted party is obtained on the first call, and that 10% of telephoning time is used on wasted calls. Note that it is not the telephoning time which can be saved that is important so much as accelerating the information exchange and hence speeding up office procedures. The problem of party accessibility is likely to become worse rather than better in future because telephone traffic is increasing all the time. Furthermore, in view of the continual increase in business travel it can be assumed that absence from the office will be more and more frequent. The supplementary services offered by the ISDN to improve accessibility, such as *completion of calls to busy subscriber, call waiting* (see Sect. 2.3.3) will significantly contribute to increased efficiency in the office. In this respect the ISDN can be said to provide secretarial services for every subscriber.

It should also be mentioned that the ISDN will enable some special types of telephone communication already available to be improved. The advantages of the ISDN in handsfree speaking have already been mentioned in Sect. 5.3.1. With respect to telephone conferences, the signaling channel provides better control and thus more efficient conferencing. The two B-channels could also be used to implement a type of artificial stereophony to facilitate identification of the speakers in conference calls. Finally, use of the *voice mail* service can be simplified by indicating voice mail system user instructions on the ISDN telephone display, thus avoiding laborious voice output for user prompting.

8.1.2 Non-Voice Communication

8.1.2.1 Significance of Service Integration

Of crucial importance for service integration is the fact that, with the widespread availability of Personal Computers, an increasing number of office desks have electronic equipment for local processing and storage of text, data and graphics, in addition to telephones (Fig. 8.1). Because of their inexpensiveness, personal computers can be cost-effectively employed for a wide range of applications, and they can also be justified in places where they are less intensively used than in typically procedure-oriented desk of an accountant, inventory controller, typist, etc. Hence personal computers are increasingly being used at general office desks supporting a mixture of office activities such as text editing, running specific user programs or preparing graphic displays, as well as performing personal tasks such as maintaining appointment diaries and private files.

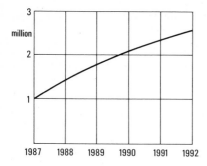

Fig. 8.1. Growth of PC-Based Office Workstations in Germany (after Diebold)

Every desk equipped with text and data processing equipment basically also needs electronic communication facilities so that processed texts can be forwarded to their destinations, databases accessed, or programs retrieved from a central program memory. Services such as text mail and videotex intensify the need for communication still further. Office desk-top equipment is therefore being provided with communications capability, with the result that text, data and facsimile communication are being increasingly used in addition to voice.

At these desks, where communication in several information types is necessary, the advantages of an integrated services network are fully realized:

- All the terminals of an office desk can be reached under *one* directory number. The dedicated terminal (or the appropriate unit of a multiservice terminal) is accessed automatically. Users do not have to contend with a multiplicity of directory numbers, possibly of different formats (e.g. telephone number and teletex subscriber address with digits and letters).
- All the communication procedures operate to a standard pattern, because only *one* network is involved. Consequently, when additional terminals are added to the office desk, learning periods are shorter and the incidence of user errors while working with the communication system is minimized. The fact that communication with a party is simultaneously possible via both 64 kbit/s channels of the ISDN basic access can be utilized to communicate in more than one information type, e.g. transmitting a facsimile or text during a telephone call. This considerably improves communication especially in the business sector.
- It is also possible to have connections simultaneously to two different destination thanks to the two 64 kbit/s channels. During a telephone call the user could be in contact with a data processing system or a database to retrieve information needed for the telephone call or to record data resulting from the telephone call immediately.
- The advantage of a standard socket for the ISDN terminals of all services should not be underestimated. Quite apart from the fact that there is no need to ensure that the right sockets are installed for the particular terminal configuration, terminals can be flexibly connected to the S-bus to meet changed arrangements within a room.

- Since the multichannel ISDN access can be implemented using a normal telephone circuit, no new wiring is required for ISDN introduction. If a desk is equipped with a second terminal besides the telephone, e.g. a Personal Computer with communication capability, it is no longer necessary to provide a second line.

8.1.2.2 Transmission Speed

Bit rate requirements for text, data and facsimile transmission in the business sector depend to a large extent on the specific applications. Fig. 8.2 contains a summary of these application-dependent bit rate requirements and relates them to the 64 kbit/s rate of the B-channel.

The B-channel can generally be regarded as adequate for direct communication between people. It also allows direct personal information exchange in case of facsimile communication, because just a few seconds are required to transmit an A4 page. The exchange of TV freeze frames is also fast enough (see Sect. 2.3.1.2). However, the discernible trend is towards sending documents by fax rather than by post, particularly in international traffic. This often involves voluminous documents for distribution in many copies. Even faster transmission would be desirable here.

However, with facsimile communication the higher speed of the network is used not only to reduce the transmission time per page, but also to increase the quality of the transmitted fax. The high-speed ISDN channel can now also send faxes with better resolution (300 and 400 dpi (dots per inch)) in an acceptable time. In addition, self-correcting transmission procedures can be used, so that errors in reproduction (e.g. line errors) are virtually eliminated. Letter-quality fax will therefore become possible for the first time.

As soon as workstations and computers become participants in the communication process, i.e. communication is actually taking place between

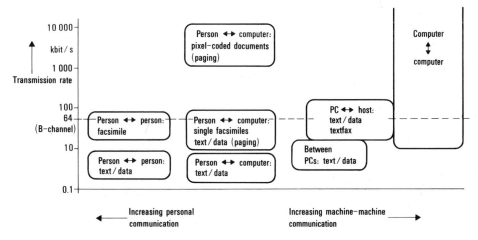

Fig. 8.2. Transmission Rate Requirements for Various Forms of Non-Voice Communication

memories, many different applications become possible, with correspondingly varied requirements in terms of transmission speed. If workstations are exchanging files with hosts or if they are remotely loading programs, the capacity of the B-channel is generally adequate. For file transfer between computers, especially in integrated operation (data, load-sharing, functional and safeguarding integration) there are cases where speeds in excess of 64 kbit/s would be desirable. The B-channel will also be too slow where the aim is to browse through pixel-coded documents stored in a remote database or archive. This application is gaining in importance with the deployment of personal computers with pixel-based screens. Assuming a frequency of two pages per second for efficient browsing and taking into account the overhead times that occur in data transmission, the resultant bit rates are in the megabits-per-second range. The same applies if large numbers of facsimiles are transmitted in batch mode in companies' internal electronic mail systems. Channels operating at some tens or hundreds of megabits per second, such as are under discussion for broadband ISDN, are more suitable for applications of this kind. Local Area Networks (LANs) and Metropolitan Area Networks (MANs) are also a possibility for local transmission (cf. Sect. 3.6.2).

At this point, some discussion of call set-up times is necessary. The call set-up time for circuit-switched connections in the public ISDN will be between 1 and 3 seconds depending on the number of exchanges involved. In conjunction with pushbutton dialing this is considerably faster than in today's telephone network but generally slower than in special data communication networks. The speed of call set-up achieved is sufficient for telephony, for occasional data exchange with hosts and messaging systems and for occasional direct transmission of text and facsimile in the user situations described in Sect. 8.1.2.1. For working at office workstations with frequent access to databases, for which the users require shorter call setup times, semipermanent and permanent links with typical activation times of 600 and 50 ms respectively can be envisaged in the ISDN.

8.1.2.3 Message Structure

As well as circuit-switched 64 kbit/s connections, the ISDN will also provide 64 kbit/s packet mode connections. Details and the advantages of the packet switching principle are given in Chap. 4. The main advantages include efficient use of trunk circuits and computer ports for interactive data communication between workstations and hosts (dialog mode).

Terminals are evolving from simple retrieval devices (such as video data terminals) into workstations having their own processing and storage systems. A change of the structure of messages exchanged between workstation and host is therefore to be expected. Whereas in the case of simple data terminals the entire transaction must be conducted directly with the host, a workstation offers the user the possibility of handling interactive work locally in most cases. Contact with the host is mainly confined to occasional transfer of large quantities of data (e.g. a text prepared locally) or of programs used at the workstation. In this situation it is also conceivable to use a circuit-switched B-channel and to set up the connection for the data transmission phases only. From the point of view of line economy, therefore, the demand for packet-switched B-channels may become less urgent as the above-mentioned terminal evolution proceeds. However, packet mode would still have the advantage of the generally shorter "call set-up time" and the ability to support several virtual connections simultaneously, permitting among other things the address-multiplex connection which is advantageous for computers.

8.2 ISDN in the Home

An increasing demand for communication facilities in the home, though not as marked as in the office environment, is nevertheless a discernible trend. This is due to a number of factors:

- As the result of increased mobility for professional reasons or to improve the quality of housing, families and circles of acquaintances are more widely scattered.
- Increasingly high-quality, low-cost electronic equipment for home entertainment, information and communication is coming onto the market (video recorders, home computers, video disk players).
- Private business transactions (banking, insurance, tax, shopping) are becoming ever more numerous and complex. At the same time there is a growing demand for information for everyday living (public transport, hours of business, events).
- Leisure time is on the increase.
- Security is becoming a growth area.

Table 8.1 shows how the potential uses of communication and information technology will develop in different spheres of private life. The ISDN may be seen as providing a sound basis for efficient, user-friendly implementation of a number of these potential applications.

The universal, multichannel access to the network is likely to be especially attractive for home users. In the home, as in the office environment, a demand will arise for communication in different information types. Videotex will become established alongside voice communication. As home computer ownership increases, so will the requirement for text and data communication.

As well as service integration, multichannel network access will be welcomed in private households where, as a rule, several people share a single line. For example, it will be possible to use the videotex service without busying out the telephone line. Unlike in the conventional double connection, in the ISDN the two 64 kbit/s channels are "bundled": in the event of a second simultaneous call attempt, the free channel is automatically selected, and each terminal can be reached via both channels, with the possibility of addressing a specific terminal if

Table 8.1. Use of Communication and Information Technology in the Home

---

Entertainment, Education

- Television
  offering a wider range of programs via cable TV, video recorders, video disk
- Electronic games
  local, but also with transmission capability
- Sound program reception/reproduction
  enhanced by digital technology
- Use of home computers for hobbies, education

Communication, Information

- Telephony
  greater convenience provided by e.g. pushbutton dialing, abbreviated dialing, name keys, alphanumeric display, by home private branch exchanges and cordless telephones
- Information retrieval
  videotex, teletext; in the more distant future via cable: cable text, image retrieval
- Text communication
  e.g. via videotex
- Remote ordering, teleshopping, banking, telesoftware via videotex

Private Office Work Using Home Computers

- Domestic bookkeeping
- Information storage
  addresses, appointments, power consumption
- Correspondence

Security and Control

- Remote control
  heating, kitchen appliances, video recorders, . . .
- Monitoring/control of power consumption
- Alarm messages:
  intrusion, fire
- Emergency calls

---

required using the equipment selection digit (direct inward dialing capability as in PABXs; see Sect. 4.3.3.2).

Messaging services, as available in some videotex systems, will be enhanced by providing an indication at the subscriber terminal that a message has been deposited in the electronic mailbox ("incoming message waiting indication", see Sects. 2.3.1.3 and 2.3.3). This function is particularly important for the private subscriber; although it is possible to notify him of received messages each time he uses videotex, this cannot be assumed to be a daily occurrence in the home environment. Moreover, the average private subscriber is likely to receive a message somewhat infrequently and so will not regularly check his mailbox. The D-channel offers the best possibility for signaling "message waiting" to the subscriber, because signals can be exchanged via this channel independently of the use of the B-channels.

The same is true if the signaling channel is used for transmission of emergency signals. This improves the effectiveness of alarm and emergency call systems.

Communication applications in the home also benefit from the high transmission rate in the ISDN access. This applies particularly to videotex. With the 64 kbit/s channel and with pages complying with the current standard [8.5, 8.6], the achievable transmission time of a few seconds for a page allows rapid browsing through texts. In addition, the loading of character sets determined by the information provider (DRCS: dynamically redefinable character sets) to prepare the display of subsequent images is then barely noticeable. The high transmission rate is also indispensable for enhancing the videotex information with photographic images. The transmission time for the latter depends on the resolution and coding. The same method whereby photographic images are represented and transmitted on videotex could also be used to exchange still images between subscribers (still image transfer service, see Sect. 2.3.1.2). Home video cameras and television sets could be used for this purpose.

The high transmission rate of the ISDN can even be used to advantage for data transmission in the home, e.g. for transmitting programs provided by computer service centers or distributors to the home computer (telesoftware).

The bus design of the ISDN access (see Sect. 4.2) also has benefits for domestic applications. The permissible line length of 100 to 150 m for the passive bus should allow the maximum permissible number of terminals to be wired up as required in a fairly large apartment or average family house (Fig. 8.3). However, if there are substantial internal traffic requirements, e.g. if it is desirable to have internal calls and external traffic at the same time, a network termination incorporating private branch exchange functions must be used. The access network will then have a star or star/bus configuration (see Sect. 4.2).

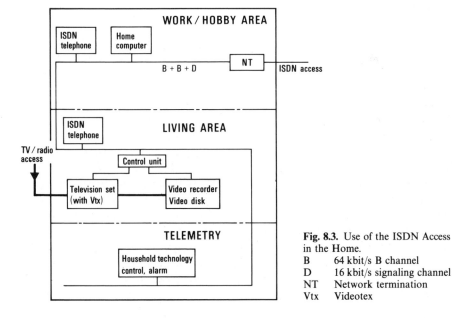

Fig. 8.3. Use of the ISDN Access in the Home.
B   64 kbit/s B channel
D   16 kbit/s signaling channel
NT   Network termination
Vtx   Videotex

As well as voice connections, data links within the home may also be required. A typical application is shown in Fig. 8.3: the television set, if equipped for videotex, can also be used as a data monitor in conjunction with the home computer. This would make it possible to use the television set in the living area to access computer programs (e.g. games) running on the home computer set up in the work area. Under control of the home computer, it would even be possible to insert television pictures from a video storage unit (video recorder, video disk player) into the program sequence.

The ISDN bus can also be used for collecting and distributing telemetry data in the home for control and monitoring purposes.

8.3 ISDN in Mobile Communication

In Sect 8.1.1 it was explained how the ISDN supplementary services for call establishment improve subscriber accessibility. These supplementary services allow a degree of subscriber mobility in the ISDN, always provided there is access to a fixed local line. The present section deals with the significance of the ISDN for mobile radio subscribers. Of the various mobile radio systems in operation today, only the *public land mobile network* is considered here, as it alone may generally be regarded as the continuation of the public wired communication network into the mobile domain. New developments in radio systems and mobile telephones are providing a very sophisticated and also cost-effective public land mobile system, and the number of subscribers has therefore increased rapidly.

At present the public land mobile network is mainly used for voice communication, e.g. using car telephones. However, with the spread of text communication in the wired network, the need to send a text to someone traveling by car will also increase. Aside from the general advantages of text communication, receiving a message in textual form has an additional advantage for mobile subscribers in a vehicle: unlike receiving a telephone call, a car driver does not have to divide his attention between driving and communicating. Many sales or service representatives in the field need up-to-date information provided by company databases, e.g. maintenance assignments. Having a data communication facility in the car saves the user the trouble of finding public telephone. Data entry while *en route* could also be important for purposes of up-to-the-minute data gathering. In conjunction with videotex, the possibility arises of using mobile data communication for retrieving the information required for car journeys: traffic reports, directions in unfamiliar localities (road and city maps), parking facilities and much else besides.

Hence the ISDN services for text and data communication also have a role to play in the public land mobile network. Moreover, the possibility of service integration offered by the ISDN has the advantage that the mobile system does not have to communicate with several different dedicated wired communication

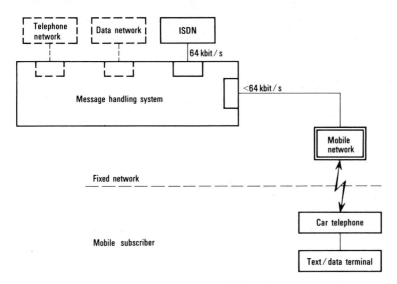

Fig. 8.4. Connection of Mobile Subscribers to the ISDN Text and Data Services via a Message Handling System

networks (e.g. the telephone network and the data network). This means that terminals installed in cars and their operation can be made more uniform. Only *one* type of interworking facility between mobile and line communication network is required.

However, it is not possible for reasons of cost and of limited frequency bandwidth resources to provide mobile subscribers generally with 64 kbit/s channels. For instance, in the European digital cellular mobile system GSM a gross bit rate of 22.8 kbit/s is envisaged. However, in mobile communication it is necessary to cope with adverse conditions of radio propagation which lead to high values of bit error ratio. Consequently, a large proportion of the gross bit rate is to be assigned to error protection; the remaining net bit rate for voice will be 13 kbit/s, and for non-voice services, 9.6 kbit/s at most. However, with the further development of mobile radio systems it might be possible to implement 64-kbit/s channels specifically for text and data communications. As long as channels with the high bit rate of 64 kbit/s are not available in the radio area, one possibility would be to provide mobile subscribers with access to the message handling system described in Sect. 2.3.1.3, which performs the appropriate rate adaption to the ISDN (Fig. 8.4). This will enable mobile subscribers to communicate with fixed subscribers to ISDN text and data communication services while those fixed subscribers maintain their normal data rate, e.g. 64 kbit/s.

Annex: CCITT Recommendations and Other Standards Relating to the ISDN

1 Introductory Remarks

Agreement has to be reached in a large number of areas to achieve the ultimate aim of communication between any two ISDN subscribers. This question is receiving the attention of a number of national, regional and international bodies.

The CCITT (Comité Consultatif International Télégraphique et Téléphonique or International Telegraph and Telephone Consultative Committee), the standardizing body of the telecommunication administrations and carriers, defines the ISDN concept as well as interfaces and signaling procedures for the ISDN in a series of recommendations; it deals equally with telecommunication services and terminals, and with transmission. The ISO (International Organization for Standardization) has prepared standards for information processing and data communication, partly in cooperation with CCITT. These standards include general basic principles of protocol architecture; since the end of 1987 this work has been the responsibility of the Joint Technical Committee ISO/IEC JTC1 (cf. Sect. 1.7 and Fig. 1.3). The IEC (International Electrotechnical Commission) produces agreements on such electrical and electromechanical questions as electrical safety, electromagnetic compatibility (EMC), and connectors. *Regional* standardizing bodies generally base their standards on those of the worldwide bodies, and also contribute to the work of the latter:

- In Europe the main regional bodies are the European Telecommunications Standards Institute ETSI for telecommunication and CEN/CENELEC for tasks mentioned in connection with ISO, IEC and JTC1. The standards of ETSI are published as European Telecommunications Standards ETS, those of CEN/CENELEC as European Standards EN. The European Computer Manufacturers Association ECMA is concerned with communication aspects of data processing systems and with ISDN PABXs (in cooperation with ETSI).
- In North America, the Standards Committee T1 – Telecommunications, develops both national standards (to be promulgated by the American National Standards Institute, ANSI) and U.S. contributions to CCITT.
- In Japan, the Telecommunication Technology Committee TTC is similar in scope and structure to T1 (but does not contribute to CCITT). In the future, the Asian ISDN Council may play a significant role in coordinating several countries.

This Annex only includes those standards that relate to the ISDN concept and the operation of voice, text and data communication systems via ISDN connections. The designation and title of each standard is given, together with a brief description of its contents.

2 CCITT Recommendations for the ISDN

The CCITT Recommendations for the ISDN listed here have mostly been published in the *Blue Book* (Geneva: International Telecommunication Union, 1989). Some however, are new or have been revised in 1990/91. They are marked correspondingly in this Annex and in the reference lists of the individual sections of this book. These new or revised recommendations are published separately; they can be ordered from the ITU.

The *I. Series* Recommendations contain all those which define the ISDN from the viewpoint of the *user*.

The *Q.* and *G. Series* Recommendations cover specifications for the *network*:

The Q. Series Recommendations relate to digital exchanges, designed in the first instance for telephony but also suitable for the ISDN, and to signaling between user and local exchange as well as between exchanges.

Recommendations relating to transmission performance and to multiplexing and transmission equipment, network synchronization and other general network aspects are covered in the G. Series. The summary below only includes those G. Recommendations directly relating to the ISDN.

The basis of a uniform "maintenance philosophy" for digital networks, including the concept of a universal Telecommunications Management Network TMN, is contained in the *M. Series*.

Recommendations on text terminals are covered in the *T. Series*.

The *X. Series* contains Recommendations for data communication including basic principles of protocol architecture (conforming to the "Open Systems Interconnection" reference model of ISO/IEC JTC1), and for interworking facilities between the ISDN and dedicated data networks.

Some of the E., Q., V. and X. Recommendations dealing with aspects affecting the ISDN user also have I.-Series designations.

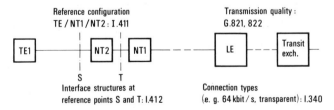

Fig. A1. Configuration of ISDN User Station and Network and Associated CCITT Recommendations with General Specifications.
TE1 ISDN terminal, NT network termination, LE local exchange (end office)

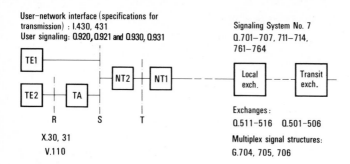

User−network interface (specifications for transmission) : I.430, 431
User signaling: Q.920, Q.921 and Q.930, Q.931

Signaling System No. 7
Q.701−707, 711−714, 761−764

Exchanges:
Q.511−516   Q.501−506

X.30, 31
V.110

Multiplex signal structures:
G.704, 705, 706

Fig. A2. Configuration of ISDN User Station and Network and Associated CCITT Recommendations with Detailed Specifications.
TE2 terminal with conventional interface (e.g. V. or X. interface), TA terminal adaptor

Figures A1 and A2 give an overview of the relation of the G., I. and Q. Recommendations to components of the ISDN.

2.1 ISDN Concepts and Principles

2.1.1 General (Blue Book, Vol. III.7)

I.110	Preamble and general structure of the I-Series recommendations for the ISDN	Overview of the I. Series recommendations
I.111	Relationship with other recommendations relevant to ISDNs	References to recommendations e.g. of the G. Series
I.112	Vocabulary of terms for ISDNs	Definitions
I.113*	Vocabulary of terms for broadband aspects of ISDN	
I.121*	Broadband aspects of ISDN	Basics of B-ISDN (Bit rates, ATM principles etc.)

2.1.2 Services (Blue Book, Vol. III.7)

I.210	Principles of telecommunication services supported by an ISDN and the means to describe them	Fundamentals of ISDN services
I.231	Circuit-mode bearer service categories	Services for data transmission etc.
I.241	Teleservices supported by an ISDN	Fully defined services (i.e. up to layer 7)
I.250	Definition of supplementary services	Basics of suppl. services (details are in I.251 to I.257)

2.1.3 Network Concept (Blue Book, Vol III.8)

I.310	ISDN – network functional principles	General information on the network concept

*Revised version (1991)

I.320	ISDN protocol reference model	Protocol reference model for ISDN based on the OSI model of ISO/IEC JTC1
I.330	ISDN numbering and addressing principles	Fundamentals of subscriber numbering and addressing of terminals
E.164* (same as I.331)	Numbering plan for the ISDN era	Subscriber numbering in the telephone network and in the ISDN
I.340	ISDN connection types	Connection types (e.g. 64 kbits/s, transparent)

2.1.4 User-Network Interfaces (Blue Book, Vol. III.8)

I.410	General aspects and principles relating to recommendations on ISDN user-network interfaces	General information on user-network interfaces
I.411	ISDN user-network interfaces – reference configurations	Arrangement of terminals, network terminations etc.
I.412	ISDN user-network interfaces – interface structures and access capabilities	Multiplex structures at user-network interfaces
I.420	Basic user-network interface	Reference to interface recommendations for basic access (B + B + D)
I.421	Primary rate user-network interface	Reference to interface recommendations for primary rate access (e.g. 30 ×B + D)
I.430	Basic user-network interface – Layer 1 specification	Specifications for layer 1
I.431	Primary rate user-network interface – Layer 1 specification	As I.430 but for primary rate access

2.1.5 Adaption of Lower User Bit Rates to 64 kbit/s, and Adaption of Existing Interfaces to the ISDN – cf. Sect. 2.8 of this Annex. (I. Recs. in Vol. III.8, X. Recs. in Vol. VIII.2 and V. 110 in Vol. VIII.1)

I.460	Multiplexing, rate adaption and support of existing interfaces	Adapting e.g. 8, 9.6, 16 kbit/s to 64 kbit/s
X. 30 (same as I.461)	Support of X.21, X.21 bis and X. 20 bis based data terminal equipments (DTEs) by an ISDN	Connection of terminals whose interface is designed for circuit-switched data networks
X. 31 (same as I.462)	Support of packet mode terminal equipment by an ISDN	Connection of terminals whose interface is designed for packet-switched data networks
V.110 (same as I.463)	Support of data terminal equipments (DTEs) with V-series type interfaces by an ISDN	Connection of terminals whose interface is designed for modems (cf. the V. recommendations in Sect. 2.8 of this Annex)
I.464*	Multiplexing, rate adaptation and support of existing interfaces for restricted 64 kbit/s transfer capability	Use of 56 kbit/s only (USA)

---

*Revised version (1991)

2.1.6 Signaling Between ISDN Subscriber Station and Local Exchange (Digital subscriber signaling system No. 1, DSS1; Blue Book, Volumes VI.10 and VI.11)

Q.920 (same as I.440)	ISDN user-network interface, data link layer – General aspects	General information on layer 2 of the user access
Q.921 (same as I.441)	ISDN user-network interface, data link layer specification	Specifications for layer 2
Q.930 (same as I.450)	ISDN user-network interface, layer 3 – General aspects	General information on layer 3 of the user access
Q.931 (same as I.451)	ISDN user net-work interface, layer 3 specification for basic call control	Specifications for layer 3
Q.932 (same as I.452)	Generic procedures for the control of ISDN supplementary services	

2.2 Signaling System No. 7 for Interexchange Signaling

2.2.1 General (Blue Book, Vol. VI.7)

Q.701 to Q.707	Signalling system No. 7 – Message transfer part (MTP)	Message transfer part

2.2.2 Control Part for Signaling Transactions (Blue Book, Vol. VI.7)

Q.711	Functional description of the signalling connection control part (SCCP)	Supplementing the message transfer part by end-to-end transport connections
Q.712	Definitions and functions of SCCP messages	Protocol elements
Q.713	SCCP format and codes	Format and coding
Q.714	Signalling connection control part procedures	

2.2.3 ISDN User Part (Blue Book, Vol. VI.8)

Q.761	Functional description of the ISDN User Part of signalling system No. 7	Application-oriented signalling functions between ISDN exchanges
Q.762	General function of messages and signals	Protocol elements
Q.763	Formats and codes	
Q.764	Signalling procedures	

2.2.4 Transaction capabilities application part (TCAP) for use in the "Intelligent Network" (Blue Book, Vol. VI.9)

Q.771	TCAP-Functional description of transaction capabilities	Objectives and architecture of transaction capabilities (TC)
Q.772	TCAP-Transaction capabilities information element definitions	Description of the message elements for transactions
Q.773	TCAP-formats and encoding	Format and encoding of transaction messages

| Q.774 | TCAP-Transaction capabilities procedures | Transaction control and component sublayer procedures for TC based on a connectionless network |

2.2.5 Operations and maintenance application part (OMAP; in Blue Book, Vol. VI.9)

| Q.795 | Operations, administration and maintenance Part | Application of S.S. No. 7 e.g. for messages between an ISDN exchange and a maintenance center |

2.3 Digital Exchanges (Blue Book, Vol. VI.5)

Recommendations for local exchanges, transit exchanges and combined exchanges.

Q.500	Introduction and field of application	Overview, including application for ISDN
Q.511 to Q.513	Exchange interfaces . . .	Interfaces to interexchange trunks, subscriber lines, and for O & M
Q.521	Exchange functions	Operational functions, e.g. for connections through the exchange
Q.522	Exchange connections	
Q.541 to Q.543	Design objectives	Performance in general
Q.551 to Q.554	Transmission characteristics	Performance at analog and digital ports

2.4 General Aspects of the Digital Network

2.4.1 Structure, Interfaces (Blue Book, Vol. III.4, except G.703/4/7)

G.701	Vocabulary of digital transmission and multiplexing, and pulse code modulation terms	Definitions
G.702	Digital hierarchy bit rates	"Plesiochronous" digital hierarchies, based on 1544 and 2048 kbit/s
G.703*	Physical/electrical characteristics of hierarchical digital interfaces	Interfaces at 64 to 155 520 kbit/s
G.704*	Synchronous frame structures used at primary and secondary hierarchical levels	Pulse frame structures with frame length 125 µs
G.707*	Synchronous hierarchical bit rates	SDH bit rates: 155, 622 and 2488 Mbit/s
G.711	Pulse code modulation of voice frequencies	Encoding per A- and µ-law

---

*Revised version (1991)

2.4.2 Transmission Performance in the Digital Network (Blue Book, Vol. III.5)

G.811	Timing requirements at the outputs of primary reference clocks suitable for plesiochronous operation of international digital links	Requirements relating to Cesium reference clocks
G.812	Timing requirements at the outputs of slave nodes	Performance of clocks in network nodes
G.821	Error performance of an international digital connection forming part of an ISDN	Bit errors in an international 64 kbit/s ISDN connection
G.822	Controlled slip rate objectives on an international digital connection	Slips in an international 64 kbit/s ISDN connection
G.823	The control of jitter and wander within digital networks which are based on the 2048 kbit/s hierarchy	Phase jitter and wander at interfaces in the network
G.824	The control of jitter and wander within digital networks which are based on the 1544 kbit/s hierarchy	See G.823

2.4.3 Network Management, Maintenance (Blue Book, Vol. IV.1)

M.30	Principles for a telecommunication management network	Network ("TMN") for management, supervision, etc.
M.36	Principles for the maintenance of ISDNs	

2.5 Basic Recommendations for Telematic Services Applicable in Principle to the ISDN (Blue Book, Vol. II.5)

F.184	Operational provisions for the international public facsimile service between subscriber stations with Group 4 facsimile machines	Telefax via data networks or ISDN
F.200	Teletex service	Definition of the service
F.300	Videotex service	Definition of the service
F.730	Service oriented requirements for telewriting applications	Basics of "telewriting"

2.6 Telematic Terminals (Blue Book, Vol. VII.3 and VII.5, except T.90)

2.6.1 Common Specifications for Different Types of Telematic Terminals

T.51	Coded character sets for Telematic services	Choice of characters for text services

T.62	Control procedures for Teletex and Group 4 facsimile services	Common control procedures (layer 5)
T.70	Network-independent basic transport service for the Telematic services	Specifications for layer 4
T.73	Document interchange protocol for the Telematic services	Specifications for layers 6 and 7
T.90*	Characteristics and protocols for terminals for telematic services in ISDN	Basic recommendation for ISDN terminals

2.6.2 Specifications for Telefax, Teletex and Videotex Terminals (Blue Book, Vol. VII.3, VII.5/6/7)

T.0 and T.2 to T.6	(Facsimile apparatus)	Basic recommendations for telefax including Group 4
T.60, T.61, T.63, T.64	(Teletex service and terminals)	Basic properties of teletex terminals; character sets for teletex (layer 6); conformity testing
T.72	Terminal capabilities for mixed mode of operation	Combined telefax and teletex
T.100	International information exchange for interactive Videotex	Recommendations T.100 and T.101 relate to basic compatibility requirements
T.101	International interworking for Vidotex service	
T.400 (Series)	(Open document architecture; document transfer and manipulation)	Uniform rules for layout and processing of documents

2.7 Specifications for Data Traffic within Bearer Services

2.7.1 Speed Classes and Service Attributes (Blue Book, Vol. VIII.2)

X.1	International user classes of service in public data networks and ISDNs	Speed classes up to 64 kbit/s
X.2	International data transmission services and optional user facilities in public data networks and ISDNs	Options for service attributes

2.7.2 Data Transfer within the framework of Open System Interconnection (Blue Book, Vol. VIII.4 and VIII.5)

X.200	Reference model of Open Systems Interconnection for CCITT applications	Essentially the same as ISO 7498
X.210	Open Systems Interconnection layer service definition conventions	Definition of the 7 layers of the "OSI model"

---

*Revised version (1991)

X.211 to X.217	(Service definitions for Open Systems Interconnection for CCITT applications)	Definition of the "services" of the individual layers
X.224 to X.227	(Protocol specifications for Open Systems Interconnection for CCITT applications)	Specifications of protocols for the individual layers

2.7.3 Frame mode/frame relaying bearer services (e.g. for general data transmission; all recommendations published separately in 1991/92)

I.122	Framework for additional packet mode bearer services	Overview, general principles
I.233.1	ISDN frame mode bearer services (FMBS) – ISDN frame relaying bearer service	
I.233.2	ISDN frame mode bearer services (FMBS) – ISDN frame switching bearer service	
I.370	Congestion management for the ISDN frame relaying bearer service	
Q.922	ISDN data link layer specification for frame mode bearer services	Layer 2 specification
Q.933	DSSI – Signaling specification for frame mode bearer services	Layer 3 specification

2.7.4 Message Handling Systems MHS ("Mailbox" Systems: Depositing, Retrieval and Format Conversion; Blue Book, Vol. VIII.7)

X.400 (same as F.400)	System and service overview	MHS functional model, MHS services, structure of the MHS protocols
X.402 X.407 to X.420	Overall architecture	Architecture of MHS Protocols of MHS, e.g. for code and format conversion (Teletex/videotex etc.)

2.8 Interfaces of Data Terminal Equipment Conforming to Standards Other than those of ISDN (Blue Book, Vol. VIII.1 for V. Recommendations and Vol. VIII.2/3 for X. Recommendations)

V.10 (same as X.26)	Electrical characteristics for unbalanced double-current interchange circuits for general use with integrated circuit equipment in the field of data communications	Unbalanced interfaces at up to 100 kbit/s (corresponds to EIA RS-423)

V.11 (same as X.27)	Electrical characteristics for balanced double-current interchange circuits for general use with integrated circuit equipment in the field of data communications	Balanced interfaces at up to 10 Mbit/s (corresponds to EIA RS-442)
V.28	Electrical characteristics for unbalanced double-current interchange circuits	For unbalanced interfaces at up to 20 kbit/s (ANSI/EIA-232)
V.24	List of definitions for interchange circuits between data terminal equipment and data circuit-terminating equipment	Basic functions of the interchange circuits (ANSI/EIA-232)
V.22, V.22 bis, V. 23 and V.26 to V.36	(Modems for telephone-type circuits)	Recommendations V.22 to V.36 (except V.24/V.25) contain interface specifications for terminals that can be connected to ISDN terminal adaptors (TA) as per V.110; modems conforming to recommendations V.22 to V.32 can be connected to a t/r ("tip and ring") interface (i.e. voice-frequency interface of a TA).
X.21	Interface between DTE and DCE for synchronous operation on public data networks	Interface for connection to circuit-switched data networks
X.21 bis	Use on public data networks of DTE which is designed for interfacing to synchronous V-series modems	Interface for connection of terminal equipment with V. interface to circuit-switched data networks
X.25	Interface between DTE and DCE for terminals operating in the packet mode and connected to public data networks by dedicated circuit	Interface for connection to packet-switched data networks
X.32	Interface between data terminal equipment (DTE) and data circuit-terminating equipment (DCE) for terminals operating in the packet mode and accessing a packet switched public data network through a public switched telephone network or an Integrated Services Digital Network or a circuit switched public data network	Access of a terminal with X.25 interface to a packet network via a circuit-switched network (e.g. an ISDN)

2.9 Recommendations on ISDN Interworking (I. Recommendations: Blue Book, Vol.III.9, except Draft I.516, scheduled for approval in 1993; Rec. X.321, X.325: Blue Book, Vol. VIII.6; Rec. X.81: Vol. VIII.3.

I.500	General structure of the ISDN interworking recommendations	Organization, scope and objectives of interworking Recommendations

I.510	Definitions and general principles for ISDN interworking	Interworking principles for – network interworking (ISDN/non-ISDN), definition of reference configurations – ISDN–ISDN interworking (to support communication between non-identical ISDN services)
I.511	ISDN-to-ISDN layer 1 internetwork interface	Specification of the layer 1 reference point Q as a common physical interface for logically different reference points K, M, N (see chapter 6) in terms of the relevant G.700–900 series Recommendations for digital networks, transmission systems and multiplexing equipment
I.515	Parameter exchange for ISDN interworking	Out-of-band or in-band parameter exchange in conjunction with call and service negotiation to establish compatibility between end points, e.g. regarding rate adaption scheme, modem type selection, etc.
I.516 (Draft)	Modem interworking arrangements for ISDN	Refers to "TA-V" and includes layer 1 IW functions and mechanisms for selecting modem type
I.520	General arrangements for network interworking between ISDNs	– Identifies the recommendations applied to ISDN–ISDN interworking (at reference point N). – Interworking functions and call negotiation (bearer capability, connection type) between two ISDNs in the case where non-identical services are supported
I.530	Network interworking between an ISDN and a Public Switched Telephone Network (PSTN)	– Interworking configurations (interexchange and within an exchange) – ISDN bearer services and connection types suitable for ISDN–PSTN interworking – Functional requirements for ISDN–PSTN interworking – Handling of non-voice calls between ISDN and PSTN subscribers (TA-t/r, TA-V.)
X.321 (same as I.540)	General arrangements for interworking between Circuit Switched Public Data Networks (CSPDNs) and Integrated Services Digital Networks (ISDNs) for the provision of data transmission	General description of interworking cases: circuit/packet switched bearer services requested on the ISDN

| X.81 | Interworking between an ISDN circuit-switched and a Circuit Switched Public Data Network (CSPDN) | Detailed interworking procedure, e.g. signaling conversion between CCS No. 7 ISUP and X.71 |
| X.325 (same as I.550) | General arrangements for interworking between Packet Switched Public Data Networks (PSPDNs) and Integrated Services Digital Networks (ISDNs) for the provision of data transmission | General description of interworking cases:
– circuit-switched bearer service on the ISDN: interworking at the network layer (single step call set-up) or by port access (two step call set-up)
– packet-switched bearer service on the ISDN |

3 Standards of ISO and ISO/IEC JTC1 for Data Communication Services Using Open Systems Interconnection (OSI)

These Standards have been published by ISO and in part (more recently) by ISO/IEC JTC1. The standards listed below refer to "Information processing systems", with further titles as given for each standard.

ISO 7498	Open Systems Interconnection – Basic reference model	Cf. Sect. 2.7.2 of this Annex
ISO 3309	Data communication – High-level data link control procedures – Frame structure	ISO 3309, 4335 and 7809 contain specifications for data protection (layer 2) by HDLC
ISO 4335	Data communications – High-level data link control procedures – Consolidation of elements of procedures	
ISO 7809	Data communication – High-level data link control procedures – Consolidation of classes of procedures	
ISO/IEC 8208	Data communication – X.25 packet level protocol for data terminal equipment	Compare X.25 in Section 2.8 of this Annex
ISO 8348	Data communications – Network service definitions	Definition of the "services" of layer 3 (corresponds to CCITT Rec. X.213, item 2.7.2 of this Annex)
ISO 8072	Open systems interconnection – Transport service definition	Definition of the "services" of layer 4 (CCITT Rec. X.214)
ISO/IEC 8073	Open systems interconnection – Connection oriented transport protocol specification	Specifications for the layer 4 protocol (CCITT Rec. X.224)
ISO 8326	Basic connection oriented session service definition	Definition of the "services" of layer 5 (CCITT Rec. X.215)
ISO 8327	Basic connection oriented session protocol specification	Specifications for the layer 5 protocol (CCITT Rec. X.225)

4 Standards of ISO and ISO/IEC JTC1 for Local Area Networks (LANs)

ISO 8802-2	Logical link control	
ISO/IEC 8802-3	Carrier sense multiple access with collision detection (CSMA/CD) access method and physical layer specifications	Access control for LAN according to "Ethernet" principle
ISO/IEC 8802-4	Token-passing bus access method and physical layer specifications	Parts 4 and 5 of ISO/IEC 8802 deal with specification for "token passing" LANs (circulating bit pattern gives authorization to transmit)
ISO/IEC 8802-5	Token ring access method and physical layer specifications	
ISO/IEC 8802-7	Slotted ring access method and physical layer specifications	Using a busy/free indicator, fixed-length "minipackets" are carried in circulating timeslots
ISO 9314-1 and 9314-2	Fiber distributed data interface (FDDI), Part 1: Token ring physical layer protocol; Part 2: Token media access control	Optical high-speed ring system with token passing access
ISO/IEC 9314-3	Part 3: Physical layer medium dependent	

5 ISO Standards for Interface Connectors

ISO 8877	Interface connector and contact assignments for ISDN basic access interface located at reference points S and T

6 European Telecommunications Standards (ETSs)

A great number of the ETSs (especially those related to ISDN) are based on CCITT recommendations. They are compatible with CCITT but avoid options. Many ETSs include detailed testing specifications (often more voluminous than the basic standard).

For CCITT recommendations of a very basic character (cf. sections 2.1.1 to 2.1.3 or 2.4.2, 2.7.1, 2.7.2 of this Annex) there are no corresponding ETSs.

In the following, the ISDN-related ETSs are listed together with the relevant CCITT recommendations. ETSs are designated ETS 300 xyz. In the following list, only the end number (xyz) is given. The ETSs listed were partly published in 1991; the rest will be published in 1992.

6.1 User-network Interfaces and Terminal Adaptation

ETS *300...*		*based on* *CCITT Rec.*
007	Support of packet mode terminal equipment by an ISDN	X.31
011	Primary rate user-network interface. Layer 1 specification and test principles	I.431
012	Basic user-network interface. Layer 1 specification and test principles	I.430
046	Primary rate access – safety and protection (Five parts: 46–1 to 46–5)	K.22
047	Basic access – safety and protection (Five parts: 47–1 to 47–5)	K.22
077	Attachment requirements for terminal adaptors to connect to an ISDN at the S/T reference point	X.30
103	Support of CCITT Recommendation X.21, X.21bis and X.20bis based Data Terminal Equipments (DTEs) by an ISDN. Synchronous and asynchronous terminal adaptation functions	X.21 X.21bis X.20bis
104	Attachment requirements for terminal equipment to connect to an ISDN using ISDN basic access. Layer 3 aspects	Q.931
126	Equipment with ISDN interface at basic and primary rate. EMC requirements	K.22
153	Attachment requirements for terminal equipment to connect to an Integrated Services Digital Network (ISDN) using ISDN basic access	I.430
156	Attachment requirements for terminal equipment to connect to an ISDN using ISDN primary rate access	I.431

6.2 Digital Subscriber Signaling System No. 1 (Generic standards, layers 2 and 3)

ETS *300...*		*based on* *CCITT Rec.*
102-1	User-network interface layer 3. Specifications for basic call control	Q.930
102-2	User-network interface layer 3. Specifications for basic call control. Specification Description Language (SDL) diagrams	Q.931
122	Generic keypad protocol for the control of supplementary services	Q.932
125	User-network interface data link layer specification. Application of CCITT Recommendations Q.920/I.440 and Q.921/I.441	Q.920, Q.921

6.3 ISDN Services

ETS *300...*		*based on* *CCITT Rec.*
048	ISDN Packet Mode Bearer Service (PMBS). ISDN Virtual Call (VC) and Permanent Virtual Circuit (PVC) bearer services provided by the B-channel of the user access: basic and primary rate	I.232 X.32, Case A

049	Packet Mode Bearer Services (PMBS). ISDN Virtual Call (VC) and Permanent Virtual Circuit (PVC) bearer services provided by the D-channel of the user access: basic and primary rate	I.232 X.31, Case B
082	3.1 kHz telephony teleservice. End-to-end compatibility	P.31, O.131, O.132
083	Circuit mode structured bearer service category usable for speech information transfer. End-to-end compatibility	G.711, O.131 O.132
084	Circuit mode structured bearer service category usable for 3.1 kHz audio information transfer. End-to-end compatibility	G.711, O.131 O.132
101	International digital audiographic teleconference	F.710
108	Circuit-mode 64 kbit/s unrestricted 8 kHz structured bearer service category. Service description	I.231.1
109	Circuit-mode 64 kbit/s 8 kHz structured bearer service category. Service description	I.231.2
110	Circuit-mode 64 kbit/s 8 kHz structured bearer service category usable for 3.1 kHz audio information transfer. Service description	I.231.3
111	Telephony 3.1 kHz teleservice. Service description	I.241.1
120	Telefax Group 4	F.184, T.563
142	Audiovisual teleservices. Video codec for audiovisual services at p* 64 kbit/s	H.261
143	Audiovisual teleservices. System for establishing communication between audiovisual terminals using digital channels up to 2048 kbit/s	H.242
144	Audiovisual teleservices. Frame structure for a 64 to 1920 kbit/s channel	H.221
145	Audiovisual teleservices. Narrowband visual telephone systems	H.242
146	Audiovisual teleservices. Frame synchronous control and indication signals for audiovisual systems	H.230

6.4 Supplementary Services

For each supplementary service, there are three ETSs, namely for stage 1 (service description), stage 2 (functional capabilities and information flows), and stage 3 (DSS1 protocol).

ETS 300. . .		based on CCITT Rec.
050	Multiple subscriber number (MSN) supplementary service, stage 1	I.251, § 2
051	MSN, stage 2	Q.81, § 2
052	MSN, stage 3	Q.951, § 2
053	Terminal portability (TP) supplementary service, stage 1	
054	TP, stage 2	Q.83, § 4
055	TP, stage 3	
056	Call waiting (CW) supplementary service, stage 1	I.253, § 1
057	CW, stage 2	Q.83, § 1
058	CW, stage 3	Q.953, § 1
059	Subaddressing (SUB) supplementary service, stage 1	I.251, § 8
060	SUB, stage 2	Q.81, § 8
061	SUB, stage 3	Q.951, § 8
062	Direct dialling in (DDI) supplementary service, stage 1	I.251, § 1
063	DDI, stage 2	Q.81, § 1
064	DDI, stage 3	Q.951, § 1
089	Calling line identification presentation (CLIP) suppl. serv., stage 1	I.251, § 3
090	Calling line identification restriction (CLIR) suppl. serv., stage 1	I.251, § 4
091	CLIP and CLIR, stage 2	Q.81, § 3

092	CLIP, stage 3	Q.951, § 3
093	CLIR, stage 3	Q.951, § 4
094	Connected line identification presentation (COLP) suppl. serv., stage 1	I.251, § 5
095	Connected line identification restriction (COLR) suppl. serv., stage 1	I.251, § 6
096	COLP and COLR, stage 2	Q.81, § 5
097	COLP, stage 3	Q.951, § 5
098	COLR, stage 3	Q.951, § 6
128	Malicious call identification (MCID) supplementary service, stage 1	I.251, § 7
129	MCID, stage 2	Q.81, § 7
130	MCID, stage 3	Q.951, § 7
136	Closed User Group (CUG) suppl. service, stage 1	I.255, § 1
137	CUG, stage 2	Q.85, § 1
138	CUG, stage 3	Q.955, § 1
139	Call hold (HOLD) supplementary service, stage 1	I..253, § 2
140	HOLD, stage 2	Q.83, § 2
141	HOLD, stage 3	Q.953, § 2
164	Meet me conference (MMC) supplementary service, stage 1	I.254, § 5
165	MMC, stage 2	
178	Advice of charge: charging information at call set-up time (AOC-S) supplem. service, stage 1	I.256, § 2a
179	Advice of charge: charging information during the call (AOC-D) supplem. service, stage 1	I.256, § 2b
180	Advice of charge: charging information at the end of the call (AOC-E) supplem. service, stage 1	I.256, § 2c
181	AOC, stage 2	Q.86, § 2
182	AOC, stage 3	Q.956, § 2
183	Conference call add-on (CONF) supplementary service, stage 1	I.254, § 1
184	CONF, stage 2	Q.84, § 1
185	CONF, stage 3	Q.954, § 1
186	Three party suppl. serv. (3 PTY), stage 1	I.254, § 2
187	3 PTY, stage 2	Q.84, § 2
188	3 PTY, stage 3	Q.954, § 2
199	Call forwarding busy (CFB) suppl. service, stage 1	I.252, § 2
203	CFB, stage 2	Q.82, § 2
200	Call forwarding unconditional suppl. service (CFU), stage 1	I.252, § 4
204	CFU, stage 2	Q.82, § 2
201	Call forwarding no reply suppl. service (CFNR), stage 1	I.252, § 3
205	CFNR, stage 2	Q.82, § 2
202	Call deflection (CD) suppl. service, stage 1	I.252, § 5
206	CD, stage 2	Q.82, § 3
208	Freephone suppl. service (FPH), stage 1	I.256, § 4
209	FPH, stage 2	Q.86, § 4
210	FPH, stage 3	Q.956, § 4

6.5 Signaling System No. 7

ETS		*based on*
300...		*CCITT Rec.*
008	Message Transfer Part (MTP) to support international interconnection	Q.701 to Q.708
009	Signaling Connection Control Part (SCCP) (Connectionless service) to support international interconnection	Q.711 to Q.714

| 121 | Application of the ISDN user part of CCITT Signalling System No. 7 for international ISDN interconnections. CCITT Recommendation Q.767 edition 3:1991 – modified | Q.767 |
| 134 | Transaction Capabilities Application Part (TCAP) | Q.771 to Q.775 |

6.6 Particular Network Functions

ETS *300...*		*based on* *CCITT Rec.*
099	Specification of the Packet Handler Access Point Interface (PHI)	(various)
100	Routing. In support of the Memorandum of Understanding	E.172

6.7 ISDN Terminals

ETS *300...*		*based on* *CCITT Rec.*
079	Syntax-based videotex. End-to-end protocols	
080	ISDN lower layer protocols for telematic terminals	T.90
081	Teletex end-to-end protocol over the ISDN	T.61/62/64
085	3.1 kHz telephony teleservice. Attachment requirements for handset terminals	P.31
087	Facsimile group 4 class 1 equipment on the ISDN. Functional specification of the equipment	T.6, T.563
112	Facsimile group 4 class 1 equipment on the ISDN. End-to-end protocols	T.503/521/563
155	Facsimile group 4 class 1 equipment on the ISDN. End-to-end protocols tests	T.64

References

Chapter 1

1.1   Armbrüster, H.: Retrieval services with broadband ISDN – Access to information and information processing. INFOR – Information systems and operational research 27 (1989), No. 4, "Integrated Services Digital Network (ISDN)", pp. 408–417.

1.2   Rosenbrock, K. H.: Hentschel, G.: ISDN Praxis, das Handbuch der neuen Sprach-, Text-, Bild-, Datenkommunikation, Abschn. 1.2. Ulm: Neue Medienges. 1988 (Erg. 1989).

1.3   Bocker, P.: Datenübertragung, Bd. 1: Grundlagen. 2. Aufl. Berlin, Heidelberg, New York, Tokyo: Springer 1983, pp. 40 et seq.

1.4   CCITT: Recommendation X.1: International user classes of service in public data networks and integrated services digital networks (ISDNs). Blue Book, Vol. VIII.2, Geneva: ITU 1989.

1.5   CCITT: Recommendation X.25: Interface between data terminal equipment (DTE) and data circuit terminating equipment (DCE) for terminals operating in the packet mode and connected to public data networks by dedicated circuit. Blue Book, Vol. VIII.2, Geneva: ITU 1989.

1.6   CCITT: Recommendation X.28: DTE/DCE interface for a start-stop mode data terminal equipment accessing the packet assembly/disassembly facility (PAD) in a public data network situated in the same country. Blue Book, Vol. VIII.2, Geneva: ITU 1989.

1.7   Gabler, H. (Hrsg.): Text- und Datenübertragungstechnik. Heidelberg: R. v. Decker, G. Schenck 1988.

1.8   Gabler, H. (Editor): Text- und Datenvermittlungstechnik. Bd. I: Leitungsvermittlungstechnik; Bd. II: Paketvermittlungstechnik. Heidelberg: R. v. Decker, G. Schenck 1987, 1988.

1.9   Gerke, P.: Neue Kommunikationsnetze. Berlin, Heidelberg, New York: Springer 1982, pp. 32–33.

1.10   Stroh, E.: "Intelligent Network" für attraktive Dienste. Fernmelde-Praxis 66 (1989) No. 4, pp. 140–147.

1.11   Frantzen, V.; Maher, A.; Eske Christensen, B.: Towards the Intelligent ISDN – Concepts, Applications, Introductory Steps. Proceedings ICIN – International Conference on Intelligent Networks, Bordeaux, 14.–17. March 1989, pp. 152–156.

1.12   Ambrosch, W. D.; Maher, A.; Sasscer, B.: The intelligent Network. Berlin, Heidelberg, New York, London, Paris, Tokyo: Springer 1989.

1.13   Clarke, C.: The Strategic Implications of Open Network Architecture. Telecommun. (1988) No. 3, pp. 41–47, 76.

1.14   Gilhooly, D.: Open Network Provision: The Real Revolution in VANS? Telecommun. (1988) No. 3, pp. 49–60.

1.15   CCITT: Recommendation M. 30: Principles for a telecommunications management network. Blue Book, Vol. IV.1, Geneva: ITU 1989.

1.16   Rosenbrock, K. H.: ISDN – A Logical Evolution of the Digital Telephone Network. Special Issue of Jahrbuch der Deutschen Bundespost 1984, Bonn 1984, pp. 478–539. Obtainable from the Federal Ministry of Posts and Telecommunications, Section 201, P.O. Box 8001, D-5300 Bonn 1. F. R. Germany.

1.17   Kahl, P.: ISDN – Das künftige Fernmeldenetz der Deutschen Bundespost. 2. Aufl. Heidelberg 1986: R. v. Decker, G. Schenck.

1.18   Armbrüster, J.: Das universelle Breitband-ISDN – Weltweite Strategien für die Realisierung. Telematica '88, Kongreßband. Stuttgart: R. Fischer 1988, pp. 193–203.

1.19   CCITT: Recommendation I.121: Broadband Aspects of ISDN. (Revised version), Geneva: ITU 1991.

1.20   Schaffer, B.: Synchronous and Asynchronous Transfer Modes in the Future Broadband ISDN. Conference Record ICC '88 – IEEE International Conference on Communications, Philadelphia, 12.–15. June 1988, pp. 1552–1558.

Chapter 2

2.1   CCITT: Recommendation X.200: Reference model of Open Systems Interconnection for CCITT applications. Blue Book, Vol. VIII.4, Geneva: ITU 1989.

2.2   CCITT: Recommendation I.430: Basic user-network interface – layer 1 specification. Blue Book, Vol. III.8,: Geneva ITU 1989.

2.3   CCITT: Recommendation I.431: Primary rate user-network interface – layer 1 specification. Blue Book, Vol. III.8, Geneva: ITU 1989.

2.4   CCITT: Recommendation Q.920: ISDN user-network interface data link layer – general aspects. Blue Book, Vol. VI.10, Geneva: ITU 1989.

2.5   CCITT: Recommendation Q.921: ISDN user-network interface data link layer specification. Blue Book, Vol. VI.10, Geneva: ITU 1989.

2.6   CCITT: Recommendation Q.930: ISDN user-network interface layer 3 – general aspects. Blue Book, Vol. VI.11, Geneva: ITU 1989.

2.7   CCITT: Recommendation Q.931: ISDN user-network interface layer 3 specification for basic call control. Blue Book, Vol. VI.11, Geneva: ITU 1989.

2.8   CCITT: Recommendation F.184: Operational provisions for the international public facsimile service between subscriber stations with group 4 facsimile machines (Telefax 4). Blue Book, Vol. II.5, Geneva: ITU 1989.

2.9   CCITT: Recommendation F.200: Teletex service. Blue Book, Vol. II.5, Geneva: ITU 1989.

2.10   CCITT: Recommendation F.220: Service requirements unique to the processable mode number one (PM.1) used within the teletex service. Blue Book, Vol. II.5, Geneva: ITU 1989.

2.11   CCITT: Recommendation F.230: Service requirements unique to the mixed mode (MM) used within the teletex service. Blue Book, Vol. II.5, Geneva: ITU 1989.

2.12   CCITT: Recommendation F.300: Videotex service. Blue Book, Vol. II.5, Geneva: ITU 1989.

2.13   CCITT: Recommendation T.6: Facsimile coding schemes and coding control functions for group 4 facsimile apparatus. Blue Book, Vol. VII.3, Geneva: ITU 1989.

2.14   CCITT: Recommendation T.60: Terminal equipment for use in the teletex service. Blue Book, Vol. VII.3, Geneva: ITU 1989.

2.15   CCITT: Recommendation T.61: Character repertoire and coded character sets for the international teletex service. Blue Book, Vol. VII.3, Geneva: ITU 1989.

2.16   CCITT: Recommendation T.62: Control procedures for teletex and group 4 facsimile services. Blue Book, Vol. VII.3, Geneva: ITU 1989.

2.17   CCITT: Recommendation T.62 bis: Control procedures for teletex and group 4 facsimile services based on Recommendations X.215/X.225. Blue Book, Vol. VII.3, Geneva: ITU 1989.

2.18   CCITT: Recommendation T.70: Network-independent basic transport service for the telematic services. Blue Book, Vol. VII.5, Geneva: ITU 1989.

2.19   CCITT: Recommendation T.90: Characteristics and protocols for terminals for telematic services in ISDN. (Revised version), Geneva: ITU 1991.

2.20   CCITT: Recommendation T.100: International information exchange for interactive videotex. Blue Book, Vol. II.5, Geneva: ITU 1989.

2.21   CCITT: Recommendation T.101: International interworking for videotex services. Blue Book, Vol. VII.5, Geneva: ITU 1989.

2.22   CCITT: Recommendations T.400 series: Open document architecture (ODA) and interchange format, document transfer and manipulation (DTAM). Blue Book, Vol. VII.6, VII.7, Geneva: ITU 1989.

2.23   CCITT: Recommendation T.501: Document application profile MM for the interchange of formated mixed mode documents. Blue Book, Vol. VII.7, Geneva: ITU 1989.

2.24   CCITT: Recommendation T.502: Document application profile PM-11 for the interchange of character content documents in processable and formatted forms (Revised version), Geneva: ITU 1991.

2.25   CCITT: Recommendation T.503: A document application profile for the interchange of group 4 facsimile documents. (Revised version), Geneva: ITU 1991.

2.26   CCITT: Recommendation T.504: Document application profile for videotex interworking. Blue Book, Vol. VII.7, Geneva: ITU 1989.

2.27   CCITT: Recommendation T.521: Communication application profile BT.0 for document bulk transfer based on the session service (according to the rules defined in T.62 bis). Blue Book, Vol. VII.7, Geneva: ITU 1989.

2.28   CCITT: Recommendation T.522: Communication application profile BT.1 for document bulk transfer. Blue Book, Vol. VII.7, Geneva: ITU 1989.

2.29   CCITT: Recommendation T.523: Communication application profile DM.1 for videotex interworking. Blue Book, Vol. VII.7, Geneva: ITU 1989.

2.30   CCITT: Recommendation T.541: Operational application profile for videotex interworking. Blue Book, Vol. VII.7, Geneva: ITU 1989.

2.31   CCITT: Recommendation T.561: Terminal characteristics for mixed mode of operation MM. Blue Book, Vol. VII.7, Geneva: ITU 1989.

2.32   CCITT: Recommendation T.562: Terminal characteristics for teletex processable mode PM.1. Blue Book, Vol. VII.7, Geneva: ITU 1989.

2.33   CCITT: Recommendation T.563: Terminal characteristics for group 4 facsimile apparatus. (Revised version), Geneva: ITU 1991.

2.34   CCITT: Recommendation T.564: Gateway characteristics for videotex interworking. Blue Book, Vol. VII.7, Geneva: ITU 1989.

2.35   CCITT: Recommendation X.25: Interface between data terminal equipment and data circuit-terminating equipment for terminals operating in the packet mode and connected to public data networks by dedicated circuit. Blue Book, Vol. VIII.2, Geneva: ITU 1989.

2.36   CCITT: Recommendation X.75: Packet switched signaling system between public networks providing data transmission services. Blue Book, Vol. VIII.3, Geneva: ITU 1989.

2.37   CCITT: Recommendation X.208: Specification of abstract syntax notation one (ASN.1). Blue Book, Vol. VIII.4, Geneva: ITU 1989.

2.38   CCITT: Recommendation X.209: Specification of basic encoding rules for abstract syntax notation one (ASN.1). Blue Book, Vol. VIII.4, Geneva: ITU 1989.

2.39   CCITT: Recommendation X.214: Transport service definition for Open Systems Interconnection for CCITT applications. Blue Book, Vol. VIII.4, Geneva: ITU 1989.

2.40   CCITT: Recommendation X.215: Session service definition for Open Systems Interconnection for CCITT applications. Blue Book, Vol. VIII.4, Geneva: ITU 1989.

2.41   CCITT: Recommendation X.216: Presentation service definition for Open Systems Interconnection for CCITT applications. Blue Book, Vol. VIII.4, Geneva: ITU 1989.

2.42   CCITT: Recommendation X.217: Association control service definition for Open Systems Interconnection for CCITT applications. Blue Book, Vol. VIII.4, Geneva: ITU 1989.

2.43   CCITT: Recommendation X.224: Transport protocol specification for Open Systems Interconnection for CCITT applications. Blue Book, Vol. VIII.5, Geneva: ITU 1989.

2.44   CCITT: Recommendation X.225: Session protocol specification for Open Systems Interconnection for CCITT applications. Blue Book, Vol. VIII.5, Geneva: ITU 1989.

2.45   CCITT: Recommendation X.226: Presentation protocol specification for Open Systems Interconnection for CCITT applications. Blue Book, Vol. VIII.5, Geneva: ITU 1989.

2.46   CCITT: Recommendation X.227: Association control protocol specification for Open Systems Interconnection for CCITT applications. Blue Book, Vol. VIII.5, Geneva: ITU 1989.

2.47   ETSI: European Telecommunications Standards ETS 300072 to 74 on videotex presentation layer data syntax. 1991.

2.48   ISO/IEC: International Standard ISO/IEC 8208: X.25 packet level protocol for data terminal equipment. Geneva: IEC, 1990.

2.49   CCITT: Recommendation I.210: Principles of telecommunication services supported by an ISDN and the means to describe them. Blue Book, Vol. III.7, Geneva: ITU 1989.

2.50   CCITT: Recommendation I.220: Common dynamic description of basic telecommunication services. Blue Book, Vol. III.7, Geneva: ITU 1989.

2.51   CCITT: Recommendation I.230: Definition of bearer service categories. Blue Book, Vol. III.7, Geneva: ITU 1989.

2.52   CCITT: Recommendation I.231: Circuit-mode bearer service categories. Blue Book, Vol. III.7, Geneva: ITU 1989.

2.53   CCITT: Recommendation I.232: Packet-mode bearer service categories. Blue Book, Vol. III.7, Geneva: ITU 1989.

2.54   CCITT: Recommendation I.240: Definition of teleservices. Blue Book, Vol. III.7, Geneva: ITU 1989.

2.55   CCITT: Recommendation I.241: Teleservices supported by an ISDN. Blue Book, Vol. III.7, Geneva: ITU 1989.

2.56   Rosenbrock, K.H.; Hentschel, G.: ISDN Praxis, das Handbuch der neuen Sprach-, Text-, Bild-, Datenkommunikation. Chapter 4 and 6. Ulm, Germany: Neue Mediengesellschaft 1988 (Supplements 1989, 1990).

2.57   CEPT: Memorandum of Understanding on the Implementation of a European ISDN Service by 1992. London: 1989.

2.58   French, J.F.: ISDN: the bottom line adds up. telcom report 12 (1989) english edition, No. 2–3, pp. 47–51.

2.59   ETSI: ETSI Technical Report ETR 010: The ETSI basic guide on the European integrated services digital network. Valbonne: ETSI 1990.

2.60   Kahl, P.: ISDN Implementation strategy of the Deutsche Bundespost Telekom. IEEE Comm. Magaz., April 1990, pp. 47–51.

2.61   Kume, Y,.: Expansion of ISDN and the impact on new media services. Telecomm. Journal, Vol. 57, 1990, pp. 395–406.

2.62   Neigh, J.L.: The Status and Evolution of ISDN in the United States. Proc. ISS '90 – International Switching Symposium, Stockholm, 28 May–1 June 1990, Vol. 2, pp. 45–54.

2.63   Siemens, Public Communication Network Group: Facts that speak for the ISDN connection. Special print, Munich 1990. Obtainable from Siemens, ÖN, P.O. Box 700073, 8000 Munich 70, Germany.

2.64   CCITT: Recommendation F.730: Service oriented requirements for telewriting applications. Blue Book, Vol. II.5, Geneva: ITU 1989.

2.65   CCITT: Recommendation T.150: Telewriting terminal equipment, parts I–IV. Blue Book, Vol. VII.5, Geneva: ITU 1989.

2.66   Liou, M.L.: Visual Telephony as an ISDN Application. IEEE Comm. Magaz., Febr. 1990, pp. 30–38.

2.67   Yamashita, M.; Kenyon, N.D.; Okubo, S.: Standardization of Audiovisual Systems in CCITT. Proc. IMAGE'COM 90, Bordeaux, 1990, pp. 42–47.

2.68   Guichard, J.; Eude, G.; Texier, N.: State of the Art in Picture Coding for Low Bit Rate Applications. Proc. IMAGE'COM 90, Bordeaux, 1990, pp. 120–125.

2.69   CCITT: Draft Recommendation F. 720: Videotelephony services general. CCITT Doc. COM I-R27, Geneva: CCITT 1991. Final version to be published by ITU by 1992.

2.70   CCITT: Draft Recommendation F.721: Videotelephony teleservice for ISDN. CCITT Doc. COM I-R27, Geneva: CCITT 1991. Final version to be published by ITU by 1992.

2.71   CCITT: Recommendation H.261: Video codec for audiovisual services at $p \times 64$ kbit/s. Geneva: ITU 1991.

2.72   CCITT: Recommendation H.221: Frame structure for a 64 to 1920 kbit/s channel in audiovisual teleservices. Geneva: ITU 1991.

2.73   CCITT: Recommendation H.230: Frame-synchronous control and indication signals for audiovisual systems. Geneva: ITU 1991.

2.74   CCITT: Recommendation H.242: System for establishing communication between audiovisual terminals using digital channels up to 2 Mbit/s. Geneva: ITU 1991.

2.75   CCITT: Recommendation H.320: Narrowband visual telephone systems and terminal equipment. Geneva: ITU 1991.

2.76   CCITT: X.400 series Recommendations: Message handling systems. Blue Book, Vol. VIII.7, Geneva: ITU 1989.

2.77   Léger, A.; Sebestyén, I.; Frislev, T.; Poulsen, H.; Gicquel, S.; Scarr, J.: Photovideotex – towards Standardized Still Picture Telecommunication Services in Europe. Proc. IMAGE'COM 90, Bordeaux, 1990, pp. 36–41.

2.78   CCITT: Recommendation I.255.1: Closed user group (CUG). Blue Book, Vol. III.7, Geneva: ITU 1989.

2.79   CCITT: Recommendation I.251.1: Direct dialling in (DDI). Blue Book, Vol. III.7, Geneva: ITU 1989.

2.80   CCITT: Recommendation I.251.2: Multiple subscriber number (MSN). Blue Book, Vol. III.7, Geneva: ITU 1989.

2.81   CCITT: Draft Recommendation I.253.3: Completion of calls to busy subscribers (CCBS). CCITT Report COM I-R39, Geneva: CCITT 1991. Final version to be published by ITU by 1993.

2.82   CCITT: Recommendation I.253.1: Call waiting (CW), Blue Book, Vol. III.7, Geneva: ITU 1989.

2.83   CCITT: Recommendation I.252.4: Call forwarding unconditional (CFU). Blue Book, Vol. III.7, Geneva: ITU 1989.

2.84   CCITT: Recommendation I.252.3: Call forwarding no reply (CFNR). Blue Book, Vol. III.7, Geneva: ITU 1989.

2.85   CCITT: Recommendation I.252.2: Call forwarding busy (CFB). Blue Book, Vol. III.7, Geneva: ITU 1989.

2.86   CCITT: Draft Recommendation I.255.5: Outgoing call barring (OCB). CCITT Report COM I-R37, Geneva: CCITT 1991. Final version to be published by ITU by 1992.

2.87   CCITT: Draft Recommendation I.256.4: International freephone service (IFS). CCITT Report COM I-R41, Geneva: CCITT 1991. Final version to be published by ITU by 1993.

2.88   CCITT: Draft Recommendation I.256.3: Reverse charging (REV). CCITT Report COM I-R37, Geneva: CCITT 1991. Final version to be published by ITU by 1992.

2.89   CCITT: Recommendation I.253.2: Call hold (CH). Blue Book, Vol. III.7, Geneva: ITU 1989.

2.90   CCITT: Recommendation I.254.1: Conference calling (CONF). Blue Book, Vol. III.7, Geneva: ITU 1989.

2.91   CCITT: Draft Recommendation I.255.2: Private numbering plan (PNP). CCITT Report COM I-R37, Geneva: CCITT 1991. Final version to be published by ITU by 1992.

2.92   CCITT: Recommendation I.256.2: Advice of charge (AOC). Blue Book, Vol. III.7, Geneva: ITU 1989.

2.93   CCITT: Recommendation I.251.3: Calling line identification presentation (CLIP). Blue Book, Vol. III.7, Geneva: ITU 1989.

2.94   CCITT: Recommendation I.251.5: Connected line identification presentation (COLP). Blue Book, Vol. III.7, Geneva: ITU 1989.

2.95   CCITT: Recommendation X.1: International user classes of service in public data networks and integrated services digital networks (ISDNs). Blue Book, Vol. VIII.2, Geneva: ITU 1989.

2.96   CCITT: Recommendation I.121: Broadband aspects of ISDN. Blue Book, Vol. III.7, Geneva: ITU 1989.

2.97   CCITT: Recommendation I.211: Broadband ISDN service aspects. Geneva: ITU 1991.

2.98   Siemens, Public Communication Network Group: The intelligent integrated broadband network – Telecommunications in the 1990s. Special print, Munich 1990. Obtainable from Siemens, ÖN, P.O. Box 700073, 8000 Munich 70, Germany.

2.99   Patterson, J.F.; Egido, C.: Three Keys to the Broadband Future: A View of Applications. IEEE Netw. Magaz., March 1990, pp. 41–47.

2.100 Dambrowski, G.H.; Estes, G.H.; Spears, D.R.; Walters, S.M.: Implications of B-ISDN Services on Network Architecture and Switching. Proc. ISS'90 – International Switching Symposium, Stockholm, 28 May–1 June 1990, Vol. 1, pp. 91–98.

2.101 Armbrüster, H.; Rothamel, H.J.: Broadband Applications and Services. Communications Technology International 1991. The Sterling Publishing Group PLC, London: 1991.

2.102 Little, T.D.C.; Ghafoor, A.: Network Considerations for Distributed Multimedia Object Composition and Communication. IEEE Netw. Magaz., Nov. 1990, pp. 32–49.

2.103 CCITT: Draft Recommendation F.811: Broadband connection oriented bearer service. CCITT Report COM I-R27/35, Geneva: CCITT 1991. Final version to be published by ITU by 1992.

2.104 CCITT: Draft Recommendation F.812: Broadband connectionless bearer service: CCITT Report COM I-R27/35, Geneva: 1991. Final version to be published by ITU by 1992.

Chapter 3

3.1   CCITT: Recommendation M.30: Principles for a telecommunications management network. Blue Book, Vol. IV.1, Geneva: ITU 1989.

3.2   Doyle, J.S.; McMahon, C.S.: The intelligent network concept. IEEE Trans. on Commun., Vol. 36 (1988), pp. 1296–1301.

3.3   Bocker, P.: Datenübertragung, Band II: Einrichtungen und Systeme. Berlin, Heidelberg, New York: Springer 1979, pp. 25 et seq.

3.4   CCITT: Recommendation F.300: Videotex Service. Blue Book, Vol. II.5, Geneva: ITU 1989.

3.5   CCITT: Recommendation F.160: General operational provisions for the international public facsimile service. Blue Book, Vol. II.5, Geneva: ITU 1989.

3.6   CCITT: X.-Series Recommendations: Data Communication networks. Blue Book, Vol. VIII.2, VIII.3 and VIII.4, Geneva: ITU 1989.

3.7   Schollmeier, G.: The user interface in the ISDN. telcom rep. 8 (1985), Special issue "Integrated Services Digital Network ISDN", pp. 22–26.

3.8   CCITT: Recommendation E.500: Traffic intensity measurement principles. (Revised version), Geneva: ITU 1992.

3.9   Bear, D.: Principles of telecommunication-traffic engineering. Stevenage, UK, and New York: Peter Peregrinus Ltd. 1980, p. 113.

3.10   CCITT: Recommendation E.521: Calculation of the number of circuits in a group carrying overflow traffic. Blue Book, Vol. II.3, Geneva: ITU 1989.

3.11   Daisenberger, G.; Reger, J.; Wegmann, G.: Traffic measurement and monitoring, an aid for planning and operating telephone exchanges and networks. telcom rep. 4 (1981) pp. 261–269.

3.12   Grabowski, K.-H.; Hagenhaus, L.: Traffic models for ISDN with integrated packet switching. 12th Intern. Teletraffic Congr. Torino 1988, Congr. Book, pp. 4.1A2.1–4.1A2.7.

3.13   Dietrich, G.: Telephone traffic model for common control investigations. 7th Internat. Teletraffic Congr. Stockholm 1973, Congr. Book, pp. 331/1–331/6.

3.14   Gimpelson, L.: Network transition strategies – analog to IDN to ISDN. Commun. internat. 1981: (June 1981), pp. 43–46; (July 1981), pp. 40–47.

3.15   Ash, G.R.: Design and Control of Networks with Dynamic Nonhierarchical Routing. IEEE Commun. Magazine, Oct. 1990, Vol. 28, pp. 30–34.

3.16   Raab, G.: Private ISDN communications systems and their interoperation with the public ISDN. telcom rep. 8 (1985), Special Issue "Integrated Services Digital Network ISDN", pp. 57–63.

3.17   Fromm, I.: Local area networks – High-speed networks for office communications. telcom rep. 5 (1982), pp. 234–239.

3.18   CCITT: Recommendation E.163: Numbering plan for the international telephone service. Blue Book, Vol. II.2, Geneva: ITU 1989. (Also contained in revised E.164.)

3.19   CCITT: Recommendation E.164: Numbering plan for the ISDN era. (Revised version), Geneva: ITU 1991.

3.20   CCITT: Recommendation E.166: Numbering plan interworking in the ISDN era. Blue Book, Vol. II.2, Geneva: ITU 1989.

3.21   CCITT: Recommendation E.165: Timetable for coordinated implementation of the full capability of the numbering plan of the ISDN era (Recommendation E.164). Blue Book, Vol. II.2, Geneva: ITU 1989.

3.22   CCITT: Economic and Technical Aspects of the Choice of Telephone Switching Systems. Handbook, Geneva: ITU 1981, pp. 54–55.

3.23   Rosenbrock, K.H.: Possible integration of telecommunication services in the digital telephone network of the Deutsche Bundespost – ISDN. telcom rep. 5 (1982) pp. 256–260.

3.24   Schramel, F.J.: Trends in digital switching and ISDN. Telecommun. J. 49 (1982), pp. 421–429.

3.25   CCITT: Recommendation G.114: Mean one-way propagation time. Blue Book, Vol. III.1, Geneva: ITU 1989.

3.26   Claus, J.: Internationale ISDN-Aktivitäten im europäischen und außereuropäischen Ausland im Vergleich zu den Aktivitäten der Deutschen Bundespost. Online '86, 9th European Congress Fair for Technical Communications, 5.–8. 2. 1986.

3.27   Orbell, A.G.: British Telecoms plans for ISDN. IEEE Internat. Conf. on Commun., Amsterdam 1984. Conf. Rec. Vol. 2, pp. 576–579.

3.28   Kuwabara, M.: Japan is making INS a reality. telephony, Oct. 24, 1983, pp. 64–81.

3.29   Fischer & Lorenz, Ovum Ltd.: European ISDN Atlas 1991. Obtainable from Fischer & Lorenz, 65 Vangede Bygade, DK-2820 Gentofte or Ovum Ltd, 7 Rothbone Street, London W1P 1AF, UK.

Chapter 4

4.1   CCITT: Recommendation I.410: General aspects and principles relating to recommendations on ISDN user-network interfaces. Blue Book, Vol. III.8, Geneva: ITU 1989.

4.2   CCITT: Recommendation I.411: ISDN user-network interfaces – reference configurations. Blue Book, Vol. III.8, Geneva: ITU 1989.

4.3 CCITT: Recommendation X.21: Interface between data terminal equipment (DTE) and data circuit-terminating equipment (DCE) for synchronous operation on public data networks. Blue Book, Vol. VIII.2, Geneva: ITU 1989.

4.4 CCITT: Recommendation X.25: Interface between data terminal equipment (DTE) and data circuit-terminating equipment (DCE) for terminals operating in the packet mode and connected to public data networks by dedicated circuit. Blue Book, Vol. VIII.2, Geneva: ITU 1989.

4.5 Seidel, H.: Der ISDN-Basisanschluß. telcom rep. 10 (1987) Special "Multiplex- und Leitungseinrichtungen", pp. 166–172.

4.6 CCITT: Recommendation X.30 (= I.461): Support of X.21, X.21 bis and X.20 bis based DTEs by an ISDN. Blue Book, Vol. VIII.2, Geneva: ITU 1989.

4.7 CCITT: Recommendation X.31 (= I.462): Support of packet mode terminal equipment by an ISDN. Blue Book, Vol. VIII.2, Geneva: ITU 1989.

4.8 CCITT: Recommendation V.110 (= I.463): Support of DTEs with V-series type interfaces by an ISDN. Blue Book, Vol. VIII.1, Geneva: ITU 1989.

4.9 CCITT: Recommendation I.330: ISDN numbering and addressing principles. Blue Book, Vol. III.8, Geneva: ITU 1989.

4.10 ISO International Standards 8802/2.2, 3, 4: Local Area Networks.

4.11 CCITT: Recommendation I.412: ISDN user-network interfaces – interface structures and access capabilities. Blue Book, Vol. III.8, Geneva: ITU 1989.

4.12 CCITT: Recommendation I.460: Multiplexing, rate adaption and support of existing interfaces. Blue Book, Vol. III.8, Geneva: ITU 1989.

4.13 CCITT: Recommendation I.464: Multiplexing, rate adaption and support of existing interfaces for restricted 64 kbit/s transfer capability. Blue Book, Vol. III.8, Geneva: ITU 1989.

4.14 CCITT: Recommendation I.430: Basic user-network interface – Layer 1 specification. Blue Book, Vol. III.8, Geneva: ITU 1989.

4.15 ISO International Standard 8877.

4.16 Bocker, P.: Datenübertragung, Band I: Grundlagen, 2. Aufl. Berlin, Heidelberg, New York, Tokyo: Springer 1983, pp. 128 et seq.

4.17 CCITT: Recommendation I.431: Primary rate user-network interface – Layer 1 specification. Blue Book, Vol. III.8, Geneva: ITU 1989.

4.18 CCITT: Recommendation G.703: Physical/electrical characteristics of hierarchical digital interfaces. (Revised version), Geneva: ITU 1991.

4.19 CCITT: Recommendation G.704: Synchronous frame structures used at primary and secondary hierarchical levels. (Revised version), Geneva: ITU 1991.

4.20 CCITT: Recommendation G.735: Characteristics of primary PCM multiplex equipment operating at 2048 kbit/s and offering synchronous digital access at 384 kbit/s and/or 64 kbit/s. Blue Book, Vol. III.4, Geneva: ITU 1989.

4.21 CCITT: Recommendation G.737: Characteristics of an external access equipment operating at 2048 kbit/s offering synchronous digital access at 384 kbit/s and/or 64 kbit/s. Blue Book, Vol. III.4, Geneva: ITU 1989.

4.22 CCITT: Recommendation I.320: ISDN protocol reference model. Blue Book, Vol. III.8, Geneva: ITU 1989.

4.23 CCITT: Recommendation I.420: Basic user-network interface. Blue Book, Vol. III.8, Geneva: ITU 1989.

4.24 CCITT: Recommendation I.421: Primary rate user-network interface. Blue Book, Vol. III.8, Geneva: ITU 1989.

4.25 CCITT: Recommendation X.200: Reference model of open systems interconnection for CCITT applications. Blue Book, Vol. VIII.4, Geneva: ITU 1989.

4.26 ISO International Standard 7498, 1984: Information processing systems – Open Systems Interconnection – Basic reference model.

4.27 CCITT: Recommendation Q.920 (= I.440): ISDN user-network interface data link layer – General aspects. Blue Book, Vol. VI.10, Geneva: ITU 1989.

4.28 CCITT: Recommendation Q.921 (= I.441): ISDN user-network interface, data link layer specification. Blue Book, Vol. VI.10, Geneva: ITU 1989.

4.29 CCITT: Recommendation Q.930 (= I.450): ISDN user-network interface layer 3 – General aspects. Blue Book, Vol. VI.11, Geneva: ITU 1989.

4.30 CCITT: Recommendation Q.931 (= I.451): ISDN user-network interface layer 3 specification for basic call control. Blue Book, Vol. VI.11, Geneva: ITU 1989.

4.31 CCITT: Recommendation E.164 (= I.331): Numbering plan for the ISDN era. (Revised version), Geneva: ITU 1991.

4.32   CCITT: Recommendation E.163: Numbering plan for the international telephone service. Blue Book, Vol. II.2, Geneva: ITU 1989. (Also contained in revised E.164.)

4.33   ISO International Standard 3309, 1984: Data communication – High level data link control procedures – Frame structure.

4.34   ISO International Standard 4335, 1987: Data Communication – High level data link control procedures – Consolidation of elements of procedures.

4.35   ISO International Standard 7809, 1984: Data Communication – Consolidation of classes of procedures.

4.36   CCITT: Recommendation X.75: Packet-switched signaling system between public networks providing data transmission services. Blue Book, Vol. VIII.3, Geneva: ITU 1989.

4.37   CCITT: Recommendation T.70: Network-independent basic transport service for the Telematic services. Blue Book, Vol. VII.5, Geneva: ITU 1989.

4.38   CCITT: Recommendation Q.932 (= I.452): Generic procedures for the control of ISDN supplementary services. Blue Book, Vol. VI.11, Geneva: ITU 1989.

4.39   CCITT: Recommendation X.1: International user classes of service in public data networks and integrated services digital networks (ISDNs). Blue Book, Vol. VIII.2, Geneva: ITU 1989.

4.40   CCITT: Recommendation V.5: Standardization of data signaling rates for synchronous data transmission in the general switched telephone network. Blue Book, Vol. VIII.1, Geneva: ITU 1989.

4.41   CCITT: Recommendation X.121: International numbering plan for public data networks. Blue Book, Vol. VIII.4, Geneva: ITU 1985.

4.42   CCITT: Recommendation V.24: List of definitions for interchange circuits between data terminal equipment (DTE) and data circuit-terminating equipment (DCE). Blue Book, Vol. VIII.1, Geneva: ITU 1989.

4.43   CCITT: Recommendation V.25: Automatic answering equipment and/or parallel automatic calling equipment on the general switched telephone network including procedures for disabling of echo control devices for both manually and automatically established calls. Blue Book, Vol. VIII.1, Geneva: ITU 1989.

4.44   CCITT: Recommendation V.25 bis: Automatic calling and/or answering equipment on the general switched telephone network (GSTN) using the 100-series interchange circuits. Blue Book, Vol. VIII.1, Geneva: ITU 1989.

4.45   v. Kienlin, A.; Klunker, J.: Packet switching and ISDN – a powerful alliance. Proc. ISS'87, Phoenix, March 15–20, 1987, pp. A.4.1.1–A.4.1.5.

4.46   CCITT: Recommendation X.2: International data transmission services and optional user facilities in public data networks and ISDNs. Blue Book, Vol. VIII.2, Geneva: ITU 1985.

4.47   CCITT: Recommendation X.32: Interface between data terminal equipment (DTE) and data circuit-terminating equipment (DCE) for terminals operating in the packet-mode and accessing a packet switched public data network through a public switched telephone network or a circuit-switched public data network. Published separately by the ITU, Geneva, in 1986: ISBN 92-61-02911-6.

4.48   CCITT: Recommendation I.232: Packet-mode bearer service categories. Blue Book, Vol. III.7, Geneva: ITU 1989.

4.49   Thomas, A.; Coudreuse, J.P.; Servel, M.: Asynchronous time division techniques: an experimental packet network integrating video communication. Proc. ISS'84, Florence, May 1984.

4.50   Bushan, Brij: Frame Relay, Fast Packet, and Packet Switching-Convergence or Coexistence? Telecommunications, December 1990, pp. 51–54.

4.51   CCITT: Recommendation Q.922: ISDN Data Link Layer Specification for Frame Mode Bearer Services. Geneva: ITU 1991.

4.52   CCITT: Recommendation I.150: B-ISDN ATM functional characteristics. (Revised version), Geneva: ITU 1992.

4.53   CCITT: Recommendation I.361: B-ISDN ATM Layer specification. (Revised version), Geneva: ITU 1992.

4.54   CCITT: Draft Recommendation I.122: Framework for providing additional packet mode bearer services. Blue Book, Vol. III.7, Geneva: ITU 1989.

4.55   CCITT: Recommendation Q.933: Digital Subscriber Signaling System No. 1 (DSS 1) – Signaling Specification for Frame Mode Bearer Services. Scheduled for approval in 1992.

4.56   CCITT: Recommendation I.370: Congestion Management for the ISDN Frame Relaying Bearer Service. Geneva: ITU 1991.

4.57   CCITT: Recommendation I.233.1: ISDN Frame Mode Bearer Services (FMBS) – ISDN Frame Relaying Bearer Service. Geneva: ITU 1991.

4.58   CCITT: Recommendation I.233.2: ISDN Frame Mode Bearer Services (FMBS) – ISDN Frame Switching Bearer Service. Geneva: ITU 1991.

4.59   CCITT: Recommendation I.121: Broadband aspects of ISDN. (Revised version), Geneva: ITU 1991.

4.60   Frantzen, V.; Händel, R.: A solid foundation for Broadband ISDN: New CCITT Recommendations for Asynchronous Transfer Mode (ATM) pave the way for Broadband ISDN. telcom rep. international 14 (1991) No. 1, pp. 24–27.

4.61   CCITT: Recommendation I.362: B-ISDN ATM Adaptation Layer (AAL) functional description. (Revised version), Geneva: ITU 1992.

4.62   CCITT: Recommendation I.363: B-ISDN ATM Adaptation Layer (AAL) specification. (Revised version), Geneva: ITU 1992.

4.63   Wiest, G.: More intelligence and flexibility for communication networks. The challenges for tomorrow's switching systems. telcom rep. international Vol. 13 (1990). No. 4, pp. 136–139.

4.64   Schaffer, B.: ATM Switching in the Developing Telecommunication Network. Proc. ISS'90, Stockholm 1990. Vol. I, pp. 105–110.

4.65   Dyczmons, W.; Franz, R.: The pick of the packet. Comparison of packet switching services for ISDN. telcom rep. international 13 (1990) No. 5–6, pp. 174–177.

Chapter 5

5.1   CCITT:Recommendation Q.932: Generic procedures for the control of ISDN supplementary services. Blue Book, Vol. VI.11, Geneva: ITU 1989

5.2   Hirschmann, P.; Wintzer, K.: Basic principles of digital subscriber sets. IEEE Trans. on Commun. 29 (1981), pp. 173–177.

5.3   Ebel, H.; Helmrich, H.: ISDN Terminals – the Basic Facts. telcom rep. 8 (1985) Special Issue "Integrated Services Digital Network ISDN", pp. 64–68.

5.4   Krafft, W.: Talking without your hands. telcom rep. 1: (1989) pp. 132–135.

5.5   ETS 300085: ISDN; 3.1 kHz telephony teleservice. Attachment requirements for handset terminals. Valbonne: ETSI 1991. (May become NET 33)

5.6   Ohmann, F. (Editor): Kommunikations-Endgeräte. Berlin, Heidelberg, New York, Tokyo: Springer 1983, Chapters 5 and 6.

5.7   CCITT: Recommendation T.90: Characteristics and protocols for terminals for telematic services in ISDN. (Revised version), Geneva: ITU 1991.

5.8   Kaufmann, P.; Saubert, L.: BITEL – Videotex Telephone T3210. telcom rep. 6 (1983), pp. 283–288.

5.9   Gilch, G.: BITEL – A New Communication Instrument. telcom rep. 6 (1983), pp. 288–291.

5.10   ISO/IEC: Doc 10918-1 – CCITT: Recommendation T.81: Digital Compression and Coding of Continuous-tone Still Images. Geneva: ISO-ITU 1992.

5.11   CCITT: Recommendations on Videotelephony for n × 64 kbit/s – Transmission and Systems Aspects (H.221, H.230, H.242, H.261, H.320). Geneva: ITU 1991.

5.12   Hölzlwimmer, H.; v. Brandt, A.; Tengler, W.: A 64 kbit/s motion compensated transform coder using vector quantization with scene adaptive codebook. IEEE Intern. Conf. on Commun., Seattle 1987. Congr. Book, Vol. 1, pp. 151–156.

5.13   Strobach, P.; Schütt, D.; Tengler, W.: Space-variant regular decomposition quadtrees in adaptive interframe coding. ICASSP, Intern. Conf. on Acoustics, Speech and Signal Processing, New York 1988, Congr. Book, M 7.8, pp. 1096–1099.

5.14   Strobach, P.: QSDPCM – A new technique in scene adaptive coding. 4th European Signal Processing Conf., Grenoble 1988. Conf. Book, pp. 1141–1144.

Chapter 6

6.1   CCITT: Recommendation G.711: Pulse code modulation (PCM) of voice frequencies. Blue Book, Vol. III.4, Geneva: ITU 1989.

6.2   Osterburg, G.D.: The new generation of the service proven digital EDS switching system. telcom rep. 5 (1982) No. 2, pp. 91–96.

6.3   Mair, E.; Hausmann, H.; Naeßl, R.: EWSP – A high performance packet switching system. Proc. ICCC '86, Munich, Sept. 15–19, pp. 359–364.

6.4   Skaperda, N.: EWSD: where it is. The basic structure and current capabilities of the EWSD digital switching system. telcom rep. 12 (1989) No. 2–3, pp. 56–60.

6.5   Raab, G.: Private ISDN communication systems and their interoperation with the public ISDN. telcom rep. 8 (1985), Special issue "Integrated Services Digital Network ISDN", pp. 57–63.

6.6   Mitterer, H.; Steigenberger, H.: EWSD – The ISDN switching system. telcom rep. 10 (1987) No. 5, pp. 235–240.

6.7   Mracek Mitchell, O.M.: Implementing ISDN in the United States. IEEE Journal on selected areas in communications, Vol. SAC-4 (1986), No. 3, pp. 398–406.

6.8   Arndt, G.; Rothamel, H.-J.: Communication services in the ISDN. telcom rep. 8 (1985), Special issue "Integrated Services Digital Network ISDN", pp. 10–15.

6.9   CCITT: Recommendation I.210: Principles of telecommunication services supported by an ISDN and the means to describe them. Blue Book, Vol. III.7, Geneva: ITU 1989.

6.10   CCITT: Recommendation I.310: ISDN – Network functional principles. Blue Book, Vol. III. 8, Geneva: ITU 1989.

6.11   CCITT: Recommendation I.340: ISDN connection types. Blue Book, Vol. VIII.8, Geneva: ITU 1989.

6.12   Schlanger, G.G.: An overview of Signaling System No. 7. IEEE Journal on selected areas in communications, Vol. SAC-4, (1986), No. 3, pp. 360–365.

6.13   Walker, M.G.: Get inside CCITT signaling system No. 7. Telephony, March 10, 1986, pp. 72–77.

6.14   Pusch, H.: Aspects of CCS 7 network configurations. Telecommunications, October 1987, pp. 240–251.

6.15   Lampe, B.; Stoll, A.: Signaling between ISDN exchanges. telcom rep. 8 (1985), Special issue "Integrated Services Digital Network ISDN", pp. 37–41.

6.16   CCITT: Recommendation Q.721: Functional description of the Signaling System No. 7 Telephone User Part (TUP). Blue Book, Vol. VI.8, Geneva: ITU 1989.

6.17   CCITT: Recommendation Q.722: General function of telephone messages and signals. Blue Book, Vol. VI.8, Geneva: ITU 1989.

6.18   CCITT: Recommendation Q.723: Formats and codes. Blue Book, Vol. VI. 8, Geneva: ITU 1989.

6.19   CCITT: Recommendation Q.724: Signaling procedures. Blue Book, Vol. VI.8, Geneva: ITU 1989.

6.20   CCITT: Recommendation Q.725: Signaling performance in the telephone application. Blue Book, Vol. VI.8, Geneva: ITU 1989.

6.21   CCITT: Recommendation X.61: Signaling System No. 7 – Data user part. Blue Book, Vol. VIII.3, Geneva: ITU 1989.

6.22   CCITT: Recommendation Q.761: Functional description of the ISDN user part of Signaling System No. 7. Blue Book, Vol. VI.8, Geneva: ITU 1989.

6.23   CCITT: Recommendation Q.762: General function of messages and signals. Blue Book, Vol. VI.8, Geneva: ITU 1989.

6.24   CCITT: Recommendation Q.763: Formats and codes. Blue Book, Vol. VI.8, Geneva: ITU 1989.

6.25   CCITT: Recommendation Q.764: Signaling procedures. Blue Book, Vol. VI.8, Geneva: ITU 1989.

6.26   CCITT: Recommendation Q.766: Performance objectives in the integrated services digital network application. Blue Book, Vol. VI.8, Geneva: ITU 1989.

6.27   CCITT: Recommendation Q.767: Application of the ISDN User Part of Signaling System No. 7 for ISDN interconnections. Geneva: ITU 1991.

6.28   Wiest, G.: More intelligence and flexibility for communication networks. The challenges for tomorrow's switching systems. telcom rep. intern. Vol. 13 (1990) No. 4, pp. 136–139.

6.29   Skaperda, N.: EWSD: where it's going. Future developments and applications of the EWSD digital switching system. telcom rep. 12 (1989) No. 2–3 pp. 60–65.

6.30   Händel, R.; Huber, M.N.: Integrated Broadband Networks. An introduction to ATM-based networks. Addison-Wesley Publishers Limited, 1991.

6.31   Frantzen, V.: Value-added services in public telecommunication networks. Proc. of ICCC '86, Munich, Sept. 15–19, pp. 44–49.

6.32   Eske-Christensen, B.; Schreier, K.; Stroh, D.: Intelligent network – a powerful basis for future services. telcom rep. 12 (1989) No. 5 pp. 148–151.

6.33   Ambrosch, W. D.; Maher, A.; Sasscer, B.: The Intelligent Network. A joint study by Bell Atlantic, IBM and Siemens. Berlin, Heidelberg, New York, London, Paris, Tokyo: Springer 1989, pp. 5–13.

6.34   Frantzen, V.; Maher, A.; Eske-Christensen, B.: Towards the Intelligent ISDN – Concepts, Applications, Introductory steps. Proc. First Intern. Conf. on Intelligent Networks, Bordeaux, March 1989, pp. 152–156.

6.35   CCITT: Recommendation Q.521: Exchange functions. Blue Book, Vol. VI.5, Geneva: ITU 1989.

6.36   CCITT: Recommendation Q.511: Exchange interfaces towards other exchanges. Blue Book, Vol. VI.5, Geneva: ITU 1989.

6.37   CCITT: Recommendation Q.512: Exchange interfaces for subscriber access. Blue Book, Vol. VI.5, Geneva: ITU 1989.

6.38   Schollmeyer, G.: The User Interface in the ISDN. telcom rep. 8 (1985), Special issue "Integrated Services Digital Network ISDN", pp. 22–27.

6.39   Neufang, K.: The EWSD digital switching network. telcom. rep. 4 (1981), Special issue "EWSD digital switching system", pp. 17–21.

6.40   CCITT: Recommendation Q.522: Digital exchange connections, signalling and ancillary functions. Blue Book, Vol. VI.5, Geneva: ITU 1989.

6.41   CCITT: Recommendation Q.701: Functional description of the message transfer part (MTP) of Signaling System No. 7. Blue Book, Vol. VI.7, Geneva: ITU 1989.

6.42   CCITT: Recommendation Q.702: Signaling data link. Blue Book, Vol. VI.7, Geneva: ITU 1989.

6.43   CCITT: Recommendation Q.703: Signaling link. Blue Book, Vol. VI.7, Geneva: ITU 1989.

6.44   CCITT: Recommendation Q.704: Signaling network functions and messages. Blue Book, Vol. VI.7, Geneva: ITU 1989.

6.45   CCITT: Recommendation Q.705: Signaling network structure. Blue Book, Vol. VI.7, Geneva: ITU 1989.

6.46   CCITT: Recommendation Q.706: Message transfer part signalling performance. Blue Book, Vol. VI.7, Geneva: ITU 1989.

6.47   CCITT: Recommendation Q.707: Testing and maintenance. Blue Book, Vol. VI.7, Geneva: ITU 1989.

6.48   CCITT: Recommendation Q.708: Numbering of international signalling point codes. Blue Book, Vol. VI.7, Geneva: ITU 1989.

6.49   Ribbeck, G.: Operation and maintenance of the EWSD system. telcom rep. 4 (1981), Special issue "EWSD digital switching system", pp. 45–50.

6.50   CCITT: Recommendation Q.542: Digital exchange design objectives – Operations and maintenance. Blue Book, Vol. VI.5, Geneva: 1989.

6.51   CCITT: Recommendation Q.513: Exchange interfaces for operations, administration and maintenance. Blue Book, Vol. VI. 5, Geneva: 1989.

6.52   CCITT: Recommendation Q.795: Operations, Maintenance and Administration Part (OMAP). Blue Book, Vol. VI.9, Geneva: ITU 1989.

6.53   CCITT: Recommendation I.601: General maintenance principles of ISDN subscriber access and subscriber installation. Blue Book, Vol. III.9, Geneva: 1989.

6.54   CCITT: Recommendation Q.940: ISDN user-network interface protocol for management – General aspects. Blue Book, Vol. VI.11, Geneva: ITU 1989.

6.55   CCITT: Recommendation M.30: Principles for a telecommunications management network. Blue Book, Vol. IV.1, Geneva: ITU 1989.

6.56   Bogler, G.; Junge, U.: The future of network management. Telecommunications management network (TMN) – a flexible standard for complex networks. telcom rep. intern. 13 (1990) No. 5–6, pp. 170–173.

6.57   Wenzel, G.: CCITT common channel signaling system No. 7 in the EWSD system. telcom rep. 4 (1981), Special issue "EWSD digital switching system", pp. 35–39.

6.58   CCITT: Recommendations Q.120–Q.139: Specifications of Signaling System No. 4. Blue Book, Vol. VI.2, Geneva: ITU 1989.

6.59   CCITT: Recommendations Q.140–Q.164: Specifications of Signaling System No. 5. Blue Book, Vol. VI.2, Geneva: ITU 1989.

6.60   CCITT: Recommendations Q.711: Functional description of the signalling connection control part. Blue Book, Vol. VI.7, Geneva: ITU 1989.

6.61   CCITT: Recommendations Q.712: Definition and functions of SCCP messages. Blue Book, Vol. VI.7, Geneva: ITU 1989.

6.62   CCITT: Recommendations Q.713: SCCP formats and codes. Blue Book, Vol. VI.7, Geneva: ITU 1989.

6.63   CCITT: Recommendations Q.714: Signaling connection control part procedures. Blue Book, Vol. VI.7, Geneva: ITU 1989.

6.64   CCITT: Recommendation Q.771: Functional description of transaction capabilities. Blue Book, Vol. VI.9, Geneva: ITU 1989.

6.65   CCITT: Recommendation Q.772: Transaction capabilities information element definitions. Blue Book, Vol. VI.9, Geneva: ITU 1989.

6.66   CCITT: Recommendation Q.773: Transaction capabilities formats and encoding. Blue Book, Vol. VI.9, Geneva: ITU 1989.

6.67   CCITT: Recommendation Q.774: Transaction capabilities procedures application part. Blue Book, Vol. VI.9, Geneva: ITU 1989.

6.68   Schromm, H.: ISDN solutions to office problems. Commun. Intern., October 1987, Siemens sponsored supplement, pp. 32–33.

6.69   (Anon.) Comprehension: Cornet protocol. COM – Siemens Magazine of Computers and Communication XV (1987) No. 6, p. 37.

6.70   Le Minh, T.; Cannon, S.: ISDN-Centrex: The EWSD approach. Proc. of Globecom '86, pp. 19.5.1–19.5.5.

6.71   Pasternak, E.J.; Schulman, S.A.: Customer control of Centrex service. Proc. ISS '87, Phoenix, March 15–20, 1887, pp. C.2.3.1–C.2.3.6.

Chapter 7

7.1   CCIT: Recommendation I.120: Integrated Services Digital Networks (ISDNs). Blue Book, Vol. III.7, Geneva: ITU 1989.

7.2   Schweizer, L.: Planning aspects of quantizing distortion in telephone networks. Telecommun. J. 48 (1981), 32–36.

7.3   CCITT: Recommendation G.701: Vocabulary of digital transmission and multiplexing, and pulse code modulation (PCM) terms. Blue Book, Vol. III.4, Geneva: ITU 1989.

7.4   CCITT: Recommendation G.711: Pulse code modulation (PCM) of voice frequencies. Blue Book, Vol. III.4, Geneva: ITU 1989.

7.5   CCITT: Recommendation G.113: Transmission impairments. Blue Book, Vol. III.4, Geneva: ITU 1989.

7.6   CCITT: Recommendation G.722: 7 kHz audio-coding within 64 kbit/s. Blue Book, Vol. III.4, Geneva: ITU 1989.

7.7   CCITT: Recommendation G.726: 40, 32, 24, 16 kbit/s Adaptive Differential Pulse Code Modulation. Geneva: ITU 1991.

7.8   CCITT: Recommendation G.733: Characteristics of primary PCM multiplex equipment operating at 1554 kbit/s. Blue Book, Vo. III.4, Geneva: ITU 1989.

7.9   CCITT: Recommendation G.732: Characteristics of primary PCM multiplex equipment operating at 2048 kbit/s, and Recommendation G.735: Characteristics of primary PCM multiplex equipment operating at 2048 kbit/s. and offering synchronous digital access at 384 kbit/s and/or 64 kbit/s. Blue Book, Vol. III.4, Geneva: ITU 1989.

7.10   CCITT: Recommendation G.707: Synchronous digital hierarchy bit rates (revised). Geneva: ITU 1991.

7.11   CCITT: Recommendation G.708: Network node interface for the synchronous digital hierarchy (revised). Geneva: ITU 1991.

7.12   CCITT: Recommendation G.709: Synchronous multiplex structure (revised). Geneva: ITU 1991.

7.13   Schweizer, L.: Transmission technology for the ISDN. telcom report 11 (1988), pp. 220–223.

7.14   CCITT: Recommendation G.703: Physical/electrical characteristics of hierarchical digital interfaces. (Revised version), Geneva: ITU 1991.

7.15   Schmidt, V.; v. Winnicki, K.: Digital transmission on balanced copper pairs. telcom rep., Vol. 10 (1987) Special Issue "Multiplexing and line transmission", pp. 137–143.

7.16   ITU Publication: Optical fibres for telecommunications. Geneva: ITU 1984. ISBN 92–61–01841–6.
7.17   telcom rep. Vol. 6 (1983), Special Issue "Optical Communications". 220 pages.
7.18   CCIR: Report 338–6: Propagation data and prediction methods required for the design of terrestrial line-of-sight systems. Recommendations of the CCIR, Vol. V. Geneva: ITU 1990.
7.19   Schweizer, L.: Performance of terrestrial and satellite 64 kbit/s paths: requirements of voice and data, and standards of the future Integrated Services Digital Network (ISDN). IEEE Internat. Conf. on Commun., Boston, 1983. Conf. Rec. Vol. 1, pp. 23–27
7.20   CCITT: Recommendation G.921: Digital sections based on the 2048 kbit/s hierarchy. Blue Book, Vol. III.5, Geneva: ITU 1989.
7.21   CCITT: Recommendation G.960: Digital section for ISDN basic rate access. Blue Book, Vol. III.5, Geneva: ITU 1989.
7.22   CCITT: Recommendation G.961: Digital transmission system on metallic local lines for ISDN basic rate access. Blue Book, Vol. III.5, Geneva: ITU 1989.
7.23   Schollmeier, G.: The user interface in the ISDN. telcom rep., Vol. 8 (1985) Special Issue "Integrated Services Digital Network ISDN", pp. 22–27.
7.24   Bocker, P.: Datenübertragung, Band I: Grundlagen, 2. Aufl. Berlin, Heidelberg, New York, Tokyo: Springer 1983, pp. 219–235.
7.25   CCITT: Recommendation G.165: Echo Cancellers. Blue Book, Vol. III.1, Geneva: ITU 1989.
7.26   American National Standard ANSI T1.601–1988: ISDN Basic Access "U" Interface – Layer 1.
7.27   telcom rep. 10 (1987) Special "Radio Communication".
7.28   CCITT: Recommendation G.704: Synchronous frame structures used at primary and secondary hierarchical levels. (Revised version), Geneva: ITU 1991.
7.29   CCITT: Recommendation G.742: Second order digital multiplex equipment operating at 8448 kbit/s and using positive justification, and Recommendation G.751: Digital multiplex equipments operating at the third order bit rate of 34368 kbit/s and the fourth order bit rate of 139624 kbit/s and using positive justification. Blue Book, Vol. III.4, Geneva: ITU 1989.
7.30   CCITT: Recommendation G.743: Second order digital multiplex equipment operating at 6312 kbit/s and using positive justification, and Recommendation G.752: Characteristics of digital multiplex equipments based on a second order bit rate of 6312 kbit/s and using positive justification. Blue Book, Vol. III.4, Geneva: ITU 1989.
7.31   CCITT: Recommendation G.811: Timing requirements at the outputs of primary reference clocks suitable for plesiochronous operation of international digital links. Blue Book, Vol. III.5, Geneva: ITU 1989.
7.32   CCITT: Recommendation G.823: The control of jitter and wander within digital networks which are based on the 2048 kbit/s hierarchy, and Recommendation G.824: The control of jitter and wander within digital networks which are based on the 1544 kbit/s hierarchy. Blue Book, Vol. III.5, Geneva: ITU 1989.
7.33   CCITT: Recommendation G.812: Timing requirements at the outputs of slave clocks suitable for plesiochronous operation of international digital links. Blue Book, Vol. III.5, Geneva: ITU 1989.
7.34   ISO International Standard 3309, 1984: Data communication – High level data link procedure — Frame structure.
7.35   CCITT: Recommendation G.821: Error performance of an international digital connection forming part of an ISDN. Blue Book, Vol. III.5, Geneva: ITU 1989.
7.36   CCITT: Recommendation X.50: Fundamental parameters of a multiplexing scheme for the international interface between synchronous data networks, and Recommendation X.51: Fundamental parameters of a multiplexing scheme for the international interface between synchronous data networks using 10-bit envelope structure. Blue Book, Vol. VIII.3, Geneva: ITU 1989.
7.37   CCITT: Recommendation X.22: Multiplex DTE/DCE interface for user classes 3–6. Blue Book, Vol. VIII.2, Geneva: ITU 1989.
7.38   CCITT: Recommendation G.822: Controlled slip rate objectives on an international digital connection. Blue Book, Vol. III.5, Geneva: ITU 1989.
7.39   CCITT: Recommendation G.114: Mean one-way propagation time. Blue Book, Vol. III.1, Geneva: ITU 1989.
7.40   CCITT: Recommendation O.171: Timing jitter measuring equipment for digital systems. Blue Book, Vol. IV.4, Geneva: ITU 1989.

7.41 CCITT: Recommendation I.340: ISDN connection types. Blue Book, Vol. III.8, Geneva: ITU 1989.

Chapter 8

8.1 Bradley, K.: The Role of Voice Store and Forward. Commun. Intern. 14 (1987), No. 7, S. 48–50.

8.2 Raab, G.: Private ISDN Communication systems and their interoperation with the Public ISDN. telcom rep. 8 (1986) Special Issue "Integrated Services Digital Network", pp. 57–63.

8.3 Auel, R.: Traffic Measurement in the EMS 1200 Communication System. telcom rep. 5 (1983), pp. 24–28.

8.4 CCITT: Recommendation T.101: International interworking for Videotex services. Blue Book, Vol. VII.5, Geneva: ITU 1989.

8.5 ETSI: European Telecommunications Standards ETS 300072 . . . 76 on Videotex presentation layer, processable data and terminal facility identifier. 1991.

Subject Index

A-law 96, 137, 165, 183, 209, 225
abbreviated dialing 25, 26
absent subscriber service 27
access control 154
access to public data services 102, 112
access types 65ff
activation 74, 197
accessibility 210
adaption of existing interfaces 223
adaption of lower user bit rates 223
additional packet switching techniques 119ff
address, ISDN 85
addressing 85
A/D (analog/digital) conversion (converter)
    44, 48, 50, 182
ADPCM (adaptive differential PCM) 131, 143,
    184, 209
adress-multiplex 213
advice of charge 25, 28
AIS (alarm indication signal) 191, 192
alarm 22, 24, 185, 191, 215, 223
    services 20, 22, 24
alignment procedure 95
American National Standards Institute (ANSI)
    220
AMI (alternate mark inversion) code 70, 72f,
    77, 193
analog/digital converter 47, 50, 182
announcements 25, 28
ANSI (American National Standards Institute)
    220
application layer see protocol layer 7
ARQ (automatic repeat request) 23, 205, 206,
    208
ATM (asynchronous transfer mode) 5, 8, 83,
    120,186,190
attenuation 32, 77, 188, 189, 190
    of wire pairs in local cables 193, 194
    rain 190
availability, full 41

B channel 6, 64ff, 78, 149, 185, 195
basic access 6, 65ff, 74, 76, 146ff, 195, 223
    electrical characteristics 69ff, 75f
    frame structure 72f
    reference configuration 67ff, 221

basic service attributes 17, 18, 24
bearer services see service, bearer
BHCA see busy hour call attempt
bit error 191, 192, 204ff, 226
bit error performance 191, 208
bit error ratio 22, 37, 58, 77, 191, 192, 205, 219
bit integrity 208
bit rate 2, 7f, 19, 28, 32, 64ff, 72, 78, 125,
    184ff, 213
bit rate adaption 95ff, 223
bit sequence independence 191, 193
block dialing 92
blocking 40f
Blue Signal 192
broadband 32f
    ISDN 32
    network 5, 8
    services 32f
broadcasting sound 1f
broadcast teletex 1
burst method 195
bus 68f
    configuration 86
busy hour 43
busy hour call attempt 42ff
B8ZS code 77, 185, 191, 193
2B/1Q code 193, 197
4B/3T code 193, 197
5B/6B code 193
7B/8B code 193

cable 188ff
call 40
call barring 25, 27
call control, distributed 170
call establishment 91
    /clearing 104
    single-step- 104ff, 105, 120, 143, 231
    two-step- 104ff, 112, 143, 231
call forwarding 25, 27
call forwarding busy 25, 27
call forwarding no reply 25, 27
call forwarding unconditional 25, 27
call hold service 25, 28
call progress signals 28
call set-up 41, 214

call waiting 25, 27, 211
calling line identification presentation 25, 28, 31
carrier 8
CCIR (Comité Consultatif International des
    Radiocommunications) 2
CCITT (Comité Consultatif International
    Télégraphique et Téléphonique) 2, 9, 220ff
CCITT Blue Book 169, 221ff, 228ff
CCITT recommendations 221ff
CCNC (common channel signaling network
    control) 169
cell approach 50f
CEN (European Committee for Standardization)
    9
CENELEC (European Committee for
    Electrotechnical Standardization) 9
centre, primary 36, 191, 202, 206
    secondary 36
    tertiary 36
centrex (central exchange) 48, 173f
CEPT (Conference Européenne des
    Administrations des Postes et des Télé-
    communications) 185
cesium-beam oscillator 202
cesium reference clock 203, 226
change of position in space 150, 151
change of service 26, 181
circuit switched connection 80f, 91
circuit switching 175
clock 201ff
clock supply 203
closed user group 25, 26, 38
coaxial cable 188, 204
codec 184
common channel signaling 138f, 158ff
common channel signaling system 155
communication, distributive 1, 8
    interactive 3
    retrieval 1, 3, 8
communication network 3f, 6ff, 34
communication protocol 11, 12, 13
communication socket 7
compatibility 7, 49
compatibility check 49, 86
compatibility information 83, 86
completion of calls to busy subscribers 25, 27
concentrator 38, 148f, 198, 202
    see also digital line unit DLU
conference calling 25, 28
congestion 44
connected line identification presentation 28
connecting cord 67f
connection, establishment of 140
connection of X. 25-terminals -access to
    PSPDN (packet switched public data
    network) services 112
    ISDN virtual circuit bearer service 110f, 112

connection of data terminal equipment
    ISDN bearer service solution 102ff
    public data network access approach 104ff
connection set-up and clear down 168
connection set-up time 37f
connection type 138,140, 208f, 221, 223
connector 76, 232
control, distributed 170
conversational services 15f, 19
coordination processor 153, 155
    copper wire 6, 38, 44, 188, 193, 197
core diameter 189
corporate ISDN networks 171ff
country code 85f
CRC 185
cross-connect 186
CSPDN (circuit switched public data network)
    96, 100, 102, 177f, 230f
customer control 166, 169

D channel 6, 11ff, 22, 64f, 80ff, 186
D channel protocol 14
data 3f
data communication 221, 228f, 231
data link layer see protocol layer 2, 224
data network 37, 47f, 95, 226, 229;
    see also PDN, PSPDN
data network
    -circuit switched 13, 14, 229
    packet switched 13, 14, 229
data processing system 173, 174
data protection 231
data transfer 227
data transfer protocol 89
data transmission 1, 3, 4, 19f, 21f, 22, 24, 37f,
    184, 205f
DCE (data circuit-terminating equipment) 229
deactivation 74f
dedicated circuits 38
dedicated network see network, dedicated
degraded minute 205
delay see signal delay
delay mode 40
dialing 62, 92, 94
digit-by-digit dialing 94
digital concentrator, remote 148
digital distribution frame 186, 191
digital hierarchy 187, 225
digital line section 192
digital line unit (DLU) 154
digital multiplex hierarchy 186
digital multiplexer 148, 187, 199
digital section 191, 195, 206
digital section, hypothetical 191
digital transmission channel 182ff
direct dialing in 25, 26, 62, 94
directory inquiry services 28

directory number 7, 39, 85, 212
distribution service 15, 16, 20, 23
distributive communication
   see communication, distributive
disturbance 204, 206
DRCS (dynamically redefinable character set)
   217
DSS1 (digital subscriber signaling system no.1)
   224, 233
DTE (data terminal equipment) 99, 223, 228f

echo 54, 195ff
echo cancelation 188, 196ff
echo suppressor 54
ECMA (European Computer Manufacturers
   Association) 220
edge emitting diode 189
EIA (Electronic Industries Association) 29, 58,
   228, 229
emergency 216
encoding law 183
end-to-end signaling 40
see also signaling, end-to-end
Erlang 41
error correction 23, 24
errored second 205, 206
establishment of connection 140
Ethernet 232
Europe 55
ETS (European Telecommunication Standard)
   220, 232ff
ETSI (European Telecommunications Standards
   Institute) 9f, 220
EWSD 155, 169
exchange 34
   local 34, 36f, 39, 52, 140, 154, 186, 196,
   202f, 221, 225
   local/transit 35
   long distance 52
   originating 142
   terminating 142
   transit 34, 39, 137f, 221, 225
exchange line 181
exchange termination ET 148f, 196
existing services 29

facsimile 1, 3f, 21, 32, 37, 132, 173, 205, 207,
   213, 227, 236
fault location 58
fax mail 20, 22
FDM (frequency division multiplex) 186, 202
feature node (FN) 144
feature node, vendor see vendor feature
   node VFN
Germany 51, 201
fiber, graded-index 189
   single-mode 189

fixed destination call 26
format conversion 228
frame, digital distribution 186, 191
frame alignment 72, 185, 200
frame alignment signal 77, 185, 200
frame mode bearer service 83
frame multiplexing 120
frame relaying 122f, 228
frame structure 72f, 185
framing bit 73
freephone service 25, 27
frequency comparison pilot 202
frequency deviation 203
frequency uncertainty 202
functional protocol 95f, 128

gateway 47, 52, 48, 50
grade of service 41
graded-index fiber 189
Great Britain 54
group processor 155

H channel 32, 64ff, 76, 78
handsfree dialing 26
handsfree speaking 26, 129
HDB3 code 193
HDLC (high-level data link control) 89, 95,
   119, 154, 149, 157, 205f, 231
HDTV (high-definition television) 23, 33
HICOM 173
hierarchical (bit rate, digital signal) 186
hierarchical level 36ff
high load 41
holding time 42ff
holdover mode 204
home computer 217
hypothetical digital section 191
hypothetical reference connection 191

Identification Code 49
IEC (International Electrotechnical
   Commission) 9, 220
IEV (International Electrotechnical
   Vocabulary) VI
implementation strategy 49ff
incoming message waiting indication 28
indication of date and time 25, 28
information 1
information transfer mode 140
inslot signaling see signaling, inslot
Integrated Digital Network IDN 138
   see also telephone network
intelligent network 5, 146, 170f
INTELSAT 53
interactive call 43
interactive service 15
interchange cable 67f

interchange circuit 228, 229
interexchange signaling 103, 224
    see also Signaling System No. 7
interface 9, 12, 63ff, 76, 146ff, 186, 191, 223,
    225, 228f
    user-network see user-network interface
interface structure 65f, 223
interference 204
intermediate service part (ISP) 166
internal traffic 60
international switching center 191, 206
interworking 31, 46, 101, 102, 221, 229f
interworking unit IWU 7, 96, 99, 103
interworking with non-ISDN networks 101
ISDN (Integrated Services Digital Network)
    6, 7ff, 11, 38, 52ff
ISDN, broadband 8, 123, 143
ISDN bearer service solution 102
    , facsimile 132
    , mixed mode 133
    , multiservice 127, 135
ISDN networking service 25, 28
ISDN number 85
ISDN signaling 83ff
ISDN subaddress 62, 85
ISDN telephone 130
ISDN terminal 127, 236
    , terminal connection unit 129
    , text 132
ISDN/PDN interworking see also
    interworking unit 101
ISO (International Organization for
    Standardization) 9, 13, 14, 62, 67, 79, 89,
    221, 227, 231f
ISPBX (ISDN private branch exchange)
    see private branch exchange, ISDN
ITSTC (Information Technology Steering
    Committee) 9f
IVDT (integrated voice and data terminal) 135

Japan 54, 183, 186
jitter 69, 191, 204, 208, 226
JTC (ISO/IEC Joint Technical Committee 1 -
    Information Technology) 9
junction, direct 36
    local tandem 36
junction network 36
justification 199, 200

LAN (local area network) 45f, 62, 173, 212,
    232
    ring system 232
LAP B (link access procedure balanced) 13, 90
LAP D (link access procedure on the
    D channel) 88
laser diode 189
layer 1...7 see protocol layer 1...7

LED (light-emitting diode) 189
line, leased 4
line code 192, 195, 197
line system 192, 195
line termination (LT) 148f, 192
line trunk group (LTG) 155
local area network see LAN
local exchange see exchange, local
local network see network, local
long distance network see network, long
    distance
loss 41f
mode 41

mail see text mail, voice mail
mailbox 5, 216, 228
mailbox service 43
maintenance 37, 221, 226
malicious call identification 25, 28
man-machine interface 128
man-machine language (MML) 156
master-slave method 201
memorandum of understanding (MOU) 54
mesochronous 200
message, end-to-end 40
    handling service (MHS) 144, 228
    handling system 216, 219, 228
    transfer part (MTP) 161f, 224, 235
    waiting indication 213
messaging services 5, 16, 20, 22, 80, 216
Metropolitan Area Network (MAN) 8
mixed mode 13, 20, 21
MMS 43 code 197
mobile radio system 218
mobile telephony 184
mode-field diameter 189
modem 37, 98, 109, 229, 230
multi address calling 25, 27
multi address dialing 38
multiblock operation 89
multifunction terminal see ISDN terminal,
    multiservice
multipath propagation 190
multiple subscriber number 25, 26
multiplexer, digital 148
multiplexing of virtual calls using the LAP
    address field 120
multimedia workstation 8
m law   100, 183, 209, 225

name key 92
national destination code 85
network 3ff, 34 see also communication
    network
    circuit-switched public data see CSPDN
    local 36f, 188, 198
    long distance 36

overlay see overlay network
packet 110ff
packed-switched public data
   see PSPDN
private see private network
public land mobile 218
subscriber line 34
telephone see telephone network, see PSTN
text and data see text and data network,
   see CSPDN, see PSPDN, see PDN
toll 37, 198
trunk 36, 149
network capabilities 138f
network data base 141, 170
network dimensioning 40ff
network hierarchy see hierarchical level
network information related supplementary
   services see supplementary services,
   network information related
network interworking 31f
see also interworking unit
network management 5
network layer see protocol layer 3 12, 80, 82
network synchronization 157f, 200f, 202, 221
network termination NT 29, 56ff, 67ff, 72ff,
   115, 127, 129, 197, 217, 221f
non-transparent connection 209
non-uniform encoding 183
North America 183, 186, 220
null NT2 see Zero NT2
numbering 48, 46, 104, 223

octet-structure 64, 199
office 210
Open Network Architecture (ONA) 5
Open Network Provision (ONP) 5
Open Systems Interconnection see OSl
operation 40
operations and maintenance application part
   (OMAP) 156, 167, 225
optical fiber 6, 143, 188f, 192, 194
OSI (Open Systems Interconnection) 11ff, 79,
   221, 223, 227f, 231
OSI reference model 11, 14, 18f, 79f, 221, 231
outslot signaling see signaling, outslot
overlay network 50ff

p data 88
p information 88f
PABX (private automatic branch exchange)
   28, 94, 144, 171ff, 202
packet handler 109ff, 110ff, 114ff, 140, 144,
   146, 152, 172
packet switched connection 82
packet switching 4f
packet switching network 4f
passive bus 60, 67ff

PCM (pulse code modulation) 73, 131, 142,
   182ff, 205f, 209, 225
   see also ADPCM
PCM code word 150, 183, 201
PDN (public data network) 99ff, 227, 229;
   see also PSPDN
performance 140, 204f, 208, 225, 226
personal computer PC 136, 211, 214
phase variation 203
physical connection-layer see protocol layer 1
ping-pong method 195
plesiochronous 198, 199, 226
plug 67
Poisson distribution 204
power feeding 75f, 192
presentation layer see protocol layer 6
primary center 36f, 191, 202, 206
primary multiplex signal 184
primary rate access 6, 48, 63, 65ff, 76ff, 89,
   146ff, 185, 187, 223
primary rate access, electrical characteristics 77
primary rate access, frame structure 78
primary reference clock 201ff
printed record of call charge 28
private branch exchange PBX 6f, 48f, 60ff,
   94f, 203, 210
private branch exchange, ISDN 48, 53, 171ff
private network 5, 45, 61f, 86
protection switching 191
protocol architecture 79, 81f, 113, 121,
   159, 166, 220
protocol layer 11f, 79
protocol layer 1 12, 67, 79, 83, 223, 230
protocol layer 2 12, 79, 83f, 90, 224, 228, 231
protocol layer 3 12, 80ff, 224, 228, 233
protocol layer 4 12
protocol layer 5 12
protocol layer 6 12
protocol layer 7 13
protocol standards 13
pseudo-ternary code 193
PSPDN (packet switched public data
   network) 99, 101, 110ff, 145, 172f, 229, 231
PSTN (public switched telephone network)
   101, 230
   see also telephone network
public data network PDN 99ff
   see also PDN
pulse code modulation (PCM) 2, 150
pulse code modulation, adaptive differential
   ADPCM 143
pulse frame 185, 198, 201, 225

quantizing distortion 182
quantizing interval 182f
quantizing noise 182f

radio relay 189f, 198
rate adaption 104, 105, 223
ready for data alignment 104, 108
redialing 25, 26
redundant binary 193
reference clock 201ff
reference configuration 221, 223
reference point 56ff, 206, 221, 230
registration of incoming calls 25, 27
remote multiplexer 198
repeater 39, 188, 192, 194
retrieval communication see communication,
    retrieval
retrieval services 15, 16, 20, 23
reverse charging 25, 27, 38
route, alternative 45
    high usage 45
    last choice 45
    next choice 45
routing, message- 161

s information 88f
S interface 56f, 66
sample 182, 183
sampling interval 183
SAPI (service access point identifier) 114ff
satellite 53, 190, 197, 205, 206
satellite link 53, 190
SCCP (signaling connection control part)
    see signaling connection control part
SDH (synchronous digital hierarchy) 186, 190,
    200, 225
secondary center 191, 202, 206
security 215
selective call forwarding 27
service 4ff, 9, 11, 222, 226, 228, 230f, 233f
    bearer 14f, 19, 20, 32, 49, 83, 107f, 138f,
    139, 222, 234
    supplementary see supplementary service
    tele- 14f, 19f, 138f, 222, 234
    value added (VAS) 144
service access point identifier SAPI 114ff
    see also SAPI
service attributes 11, 17f
service control point SCP 144, 145, 152, 170
service intercommunication 31f
service integration 4, 44, 211, 213, 215
service management system SMS 170f
service module 144
service specific network 86
service switching point SSP 170f
services via D channel 20, 24
SES (severely errored second) 191, 205
session layer see protocol layer 5
short-haul (network, system) 188, 198
signal 1f

signal delay 54, 207
signaling, end-to-end 142, 164f, 167
    inslot 80, 102
    ISDN- 138
    link-by-link- 164ff
    outslot 80, 102, 119
signaling channel 64ff, 150
    see also common channel signaling
signaling connection control part SCCP
    165f, 224, 235
signaling conversion 100, 231
signaling converter 47
signaling message handling 161
signaling network 159
signaling network management 162
signaling point SP 162
Signaling System no. 7 39, 47, 50, 158ff
    186, 205, 207, 222, 224
signaling transfer point STP 161, 163
single-mode fiber 189
slip 201, 203, 206f
SLMD (subscriber line module digital) 154
socket 67
space stage 150
speech transmission 184, 205, 206
standardization 9f
station, user see user station
still image 137
still image transfer service 20, 22
stimulus protocol 95ff, 128
STM (synchronous transfer mode) 8
stuffing 199
subaddress 49, 62, 85
subrate channel 64
subrate stream 107
subscriber access 146f
subscriber line 6
subscriber line 6, 34, 38ff, 40, 44, 50,
    56ff, 146f, 187, 193f, 197f
subscriber number 49f, 85
supplementary services 7, 9, 17f, 24ff, 43,
    95, 138, 222, 234f
    access related 18, 25, 26
    connection related 18, 25, 26
    network information related 18, 25, 28
    switching, circuit 37, 100, 140, 141, 143
    packet 37, 100, 110ff, 140, 141
    time division 150f
switching network 142, 145, 152
synchronization 157, 200f, 221
    network- 157, 200f, 221

T interface 56f, 64, 66
t/r (tip and ring) interface 29, 100, 109, 229,
    230
technology 4
telecontrol 4, 20, 22, 24

telecommunication see communication
telecommunication service see service
telefax 20, 21
telefax service 37f, 43
telegraphy 1
telemetry 80
telephone channel 2
telephone network 3, 6f, 11, 20, 29, 31, 36ff,
    46ff, 85, 99, 108f, 145, 152, 182,193, 229
    see also PSTN
telephony 1ff, 20, 43, 44, 236
telescript 3
teleservice see service, tele-
teletex 3, 5, 20, 21, 31, 43, 42, 173, 205,
    227, 236
television 1, 2, 8, 38
telewriting 3, 20, 21
telex 3f, 5
terminal 6f, 39, 94ff, 127ff, 221, 226f, 236
    see also terminal equipment;
    see also ISDN terminal
terminal adaptor TA 31, 56ff, 100, 101f,
    110f, 222
terminal endpoint identifier TEI 87, 114ff
terminal equipment 56ff, 101, 229
terminal equipment configuration 67, 77, 86f, 94
terminal equipment connection cord 68
terminal selection 94
tertiary center 191, 202, 206
test loop 40, 58
text 1ff
text and data network 4, 37; see also PDN,
    PSPDN 37
text mail 20, 22
text transmission 13, 14, 21, 205, 207
textfax 133f, 205
tie line 178f
tie line traffic 178f
time division 2, 72, 77, 149f, 195
time slot 151, 185
time slot change 150
time stage 150
time-division multiplex (TDM) system 150
tip and ring interface see t/r interface
token 232
toll network see network, toll
toll ticketing 28
traffic channel 64f
traffic intensity 42f
traffic routing 44f, 52f
traffic volume 43f
transaction capabilities (TC) 171, 224f
transaction capabilities application part
    (TCAP) 171, 224f, 235
transit exchange 34, 39, 45, 221, 225
transit traffic 45
transmission channel 183

transmission medium 187, 189
transmission performance 204, 226
transmission rate 42, 196
transmission speed 113
transparent connection type 208
transport layer see protocol layer 4
transversal filter 197
tributary 186
trunk 36
    direct 35
    end office toll 35
    intertoll 35
    local 35
    long distance 35
trunk, toll connecting 35
trunk access 149
trunk circuit 35, 192
trunk group 35, 40f, 44, 53
trunk junction 36
trunk network see network, trunk
T1 system 185
T1-Telecommunications 220

United Kingdom 55
USA 27, 37, 86, 134, 184, 192,
    197, 223
user 6f, 11
user class 4, 30, 105f, 227
user part, data (DUP) 142
    ISDN (ISUP) 142, 155, 162,
    224,235
    telephone (TUP) 142, 155, 164, 166
user signaling 78ff
user station 6, 39, 56f, 202, 221
    see also reference configuration
user-network interface 6f, 9, 63f, 67, 76, 98ff,
    222f, 233
user-network signaling 98ff, 153ff
user-user information 86
user-user signaling 79, 95

V. interface 98, 144f, 108
V. recommendations 228
V. 24 interface 95, 101, 109, 229
V interface (of exchange) 148
value-added service see service, value added
vendor feature node VFN 144, 145, 152
video, full-motion 137
video conferencing 8, 32, 33
videography 1
videophone service 20, 22
videotelephony 1, 8, 33, 137
videotex 1, 3f, 20, 23, 33, 212, 216f, 227, 236
videotex service 30, 37, 43, 44
virtual call 112, 114ff, 119
virtual channel 32
virtual circuit bearer service 110

virtual connection 112
virtual private network 5
vocabulary 222, 225
voice mail 20, 22, 212

wakening call 27
wander 208, 226

window size 89
workstation 8, 136, 213f

X.21 interface 100, 229
X.25 interface 100, 110ff, 229

zero NT2   57,60,84

DATE DUE

✓